Y0-AAF-610

Geometric Algebra over Local Rings

MONOGRAPHS AND TEXTBOOKS IN PURE AND APPLIED MATHEMATICS

1. K. YANO. Integral Formulas in Riemannian Geometry (1970)
2. S. KOBAYASHI. Hyperbolic Manifolds and Holomorphic Mappings (1970)
3. V. S. VLADIMIROV. Equations of Mathematical Physics (A. Jeffrey, editor; A. Littlewood, translator) (1970)
4. B. N. PSHENICHNYI. Necessary Conditions for an Extremum (L. Neustadt, translation editor; K. Makowski, translator) (1971)
5. L. NARICI, E. BECKENSTEIN, and G. BACHMAN. Functional Analysis and Valuation Theory (1971)
6. D. S. PASSMAN. Infinite Group Rings (1971)
7. L. DORNHOFF. Group Representation Theory (in two parts). Part A: Ordinary Representation Theory. Part B: Modular Representation Theory (1971, 1972)
8. W. BOOTHBY and G. L. WEISS (eds.). Symmetric Spaces: Short Courses Presented at Washington University (1972)
9. Y. MATSUSHIMA. Differentiable Manifolds (E. T. Kobayashi, translator) (1972)
10. L. E. WARD, JR. Topology: An Outline for a First Course (1972)
11. A. BABAKHANIAN. Cohomological Methods in Group Theory (1972)
12. R. GILMER. Multiplicative Ideal Theory (1972)
13. J. YEH. Stochastic Processes and the Wiener Integral (1973)
14. J. BARROS-NETO. Introduction to the Theory of Distributions (1973)
15. R. LARSEN. Functional Analysis: An Introduction (1973)
16. K. YANO and S. ISHIHARA. Tangent and Cotangent Bundles: Differential Geometry (1973)
17. C. PROCESI. Rings with Polynomial Identities (1973)
18. R. HERMANN. Geometry, Physics, and Systems (1973)
19. N. R. WALLACH. Harmonic Analysis on Homogeneous Spaces (1973)
20. J. DIEUDONNÉ. Introduction to the Theory of Formal Groups (1973)
21. I. VAISMAN. Cohomology and Differential Forms (1973)
22. B.-Y. CHEN. Geometry of Submanifolds (1973)
23. M. MARCUS. Finite Dimensional Multilinear Algebra (in two parts) (1973, 1975)
24. R. LARSEN. Banach Algebras: An Introduction (1973)
25. R. O. KUJALA and A. L. VITTER (eds). Value Distribution Theory: Part A; Part B. Deficit and Bezout Estimates by Wilhelm Stoll (1973)
26. K. B. STOLARSKY. Algebraic Numbers and Diophantine Approximation (1974)
27. A. R. MAGID. The Separable Galois Theory of Commutative Rings (1974)
28. B. R. MCDONALD. Finite Rings with Identity (1974)
29. I. SATAKE. Linear Algebra (S. Koh, T. Akiba, and S. Ihara, translators) (1975)
30. J. S. GOLAN. Localization of Noncommutative Rings (1975)
31. G. KLAMBAUER. Mathematical Analysis (1975)
32. M. K. AGOSTON. Algebraic Topology: A First Course (1976)
33. K. R. GOODEARL. Ring Theory: Nonsingular Rings and Modules (1976)
34. L. E. MANSFIELD. Linear Algebra with Geometric Applications (1976)
35. N.J. PULLMAN. Matrix Theory and its Applications: Selected Topics (1976)
36. B. R. MCDONALD. Geometric Algebra Over Local Rings (1976)

Geometric Algebra over Local Rings

Bernard R. McDonald
Department of Mathematics
The University of Oklahoma
Norman, Oklahoma

MARCEL DEKKER, INC. New York and Basel

MARCEL DEKKER, INC.

270 Madison Avenue, New York, New York 10016

LIBRARY OF CONGRESS CATALOG CARD NUMBER: 76-41623

ISBN: 0-8247-6528-1

Current printing (last digit):
10 9 8 7 6 5 4 3 2 1

PRINTED IN THE UNITED STATES OF AMERICA

PREFACE

This monograph arose from lectures on topics in the automorphism and normal subgroup theory of classical linear groups over local rings at The University of Oklahoma. We hope the monograph will provide an introduction to the area of geometric algebra over commutative rings.

The boundaries of the subject, that is, "What is geometric algebra?" were established by Artin [1] in 1957 and Dieudonné [63] in 1951. These classic studies described the structure theory, actions and transitivity, normal subgroups, commutators, and automorphisms of the classical linear groups, e.g., the general linear, symplectic, orthogonal, unitary groups and their respective commutator subgroups, over division rings and fields.

Current research may be roughly classified into two often intertwining studies.

First, there exists a commutative algebraic or ring theoretic theory where the field is replaced by a ring and the vector space by a finitely generated projective module. A continuing investigation by many scholars over the last 20 years has charted the evolution of the classical setting into a stable form for the above context. Indeed, the K-theoretic groups K_0 , K_1 and K_2 may be thought of as being measures of and rooted in, respectively, the theory and existence of a basis, the fashion in which the elementary transvections sit in the general linear group, and the generators and relations which characterize the group of elementary transvections relative to a basis.

Second, there is a context where the field is replaced by a ring of arithmetic type, e.g., a Dedekind domain or a suitable subring of a field, and the vector space by a subset of the space which forms a lattice or a bounded submodule over the domain. While the first case is a study of projective modules and their linear algebra, this second setting embraces hard and often deep number theoretic properties of domains and their quotient fields.

We view this monograph as an introduction to the former study in its commutative form, i.e., geometric algebra over a commutative ring.

The choice of a local scalar ring was made for several reasons: (a) Undoubtedly foremost was the consideration that, for local rings, the theory is relatively complete and understood, thus deserving a unified comprehensive exposition. (b) The local ring context lends itself to introduction since, on one hand, often a parody of standard linear algebra achieves the desired results; while, on the other hand, difficulties occurring in general cases often occur for local rings, e.g., two lines may intersect at more than one point, a vector may not lie in any line, or centralizer arguments fail due to the possible presence of zero divisors. (c) The knowledge of local behavior has repeatedly proven itself useful in obtaining general results in commutative algebra.

We have attempted to maintain a level of presentation sensitive to a desire for an introduction to basic techniques, concepts and results in the subject. Consequently, we examine in detail only the general linear, symplectic and orthogonal groups. Only passing comments are made concerning groups of collinear transformations, the unitary groups and the classical projective groups. Further, only minimal introduction is given to the classification problem of symmetric inner product spaces. Each chapter terminates with a brief historical comment and discussion of current active directions of research.

This monograph was designed so that by and large the chapters and basic subdivisions may be read independently. Each of the main chapters deals with three themes: (a) a description of generators, transitivity properties of the group on the space and characterizations of the commutator subgroup, (b) a description of normal subgroups as sandwiched between congruence subgroups, (c) a classification of the automorphisms of the group. The interdependence of topics is given by the following diagram.

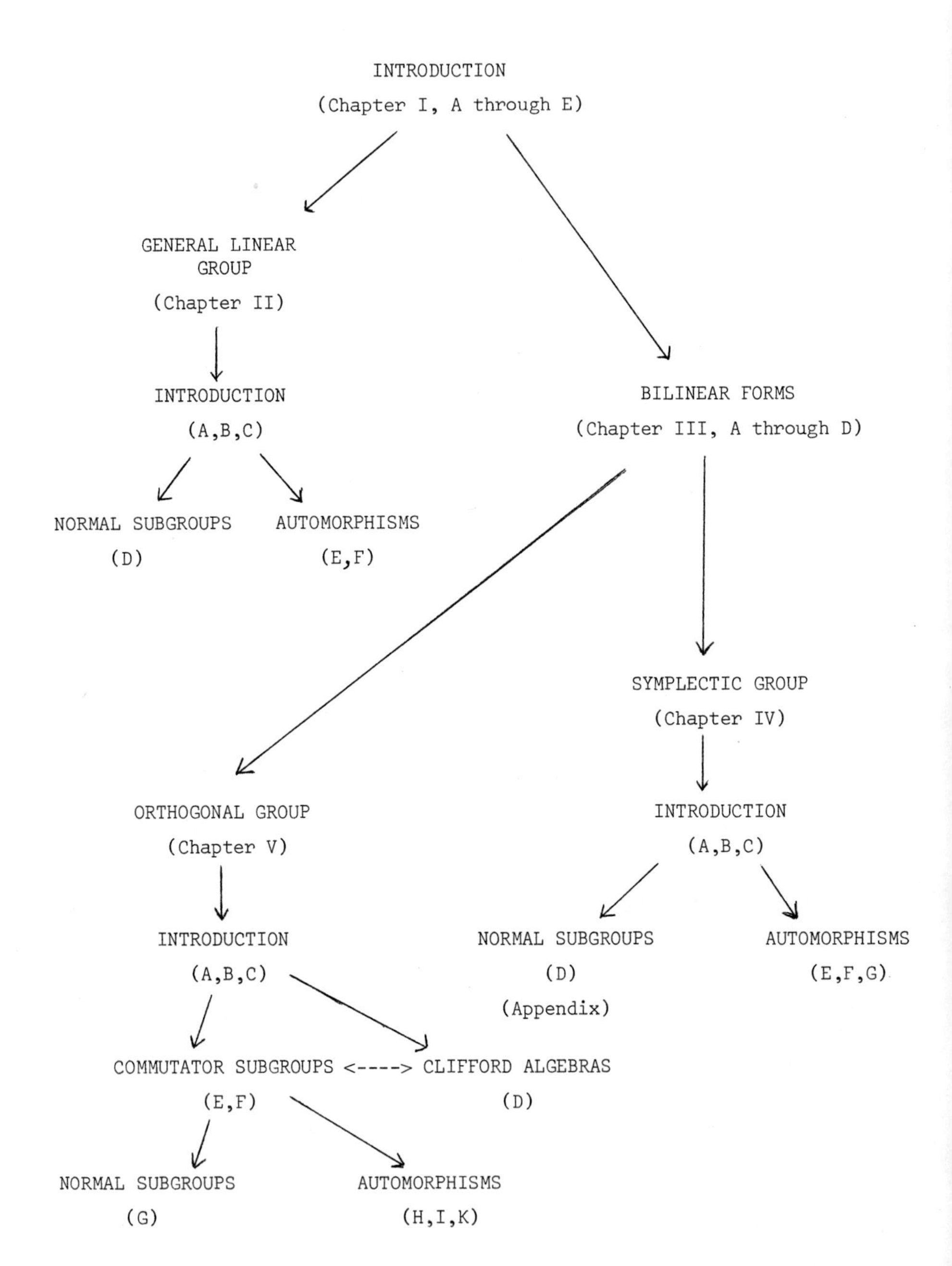
INTRODUCTION
(Chapter I, A through E)
GENERAL LINEAR GROUP
(Chapter II)
INTRODUCTION
(A,B,C)
NORMAL SUBGROUPS
(D)
AUTOMORPHISMS
(E,F)
BILINEAR FORMS
(Chapter III, A through D)
SYMPLECTIC GROUP
(Chapter IV)
INTRODUCTION
(A,B,C)
NORMAL SUBGROUPS
(D)
(Appendix)
AUTOMORPHISMS
(E,F,G)
ORTHOGONAL GROUP
(Chapter V)
INTRODUCTION
(A,B,C)
COMMUTATOR SUBGROUPS <----> CLIFFORD ALGEBRAS
(E,F)
(D)
NORMAL SUBGROUPS
(G)
AUTOMORPHISMS
(H,I,K)

Now comes the very pleasant task of expressing my appreciation to the individuals and organizations which made the writing of this monograph possible. Most of the final draft was written during the summer of 1974 while I was supported by a summer fellowship from the Arts and Sciences College and the Graduate College of The University of Oklahoma and during the following academic year at The Pennsylvania State University. I enjoyed having many enlightening discussions with Tim O'Meara, Alex Hahn, Don James, Roger Ware, Jim Pomfret and Leon McQueen. Jeanna Moore and Robb Koether carefully read the entire manuscript and offered many corrections and improvements. Trish Abolins cheerfully typed and retyped the manuscript, correcting not only my grammar and spelling, but occasionally my mathematics.

To the memory of my mother.

CONTENTS

Geometric Algebra over Local Rings

I. LOCAL RINGS AND THEIR SPACES

(A) RINGS AND MODULES - TERMINOLOGY

Let R be an associative ring with 1. An Abelian group M is called a *(left) R-module* if there is a map $R \times M \to M$, $\langle r,m \rangle \to rm$, satisfying

(a) $1m = m$

(b) $(r + s)m = rm + sm$

(c) $(rs)m = r(sm)$

(d) $r(m + n) = rm + rn$

for all r,s in R and m,n in M. A *right R-module* is defined similarly. For convenience, we refer to a left R-module as an R-module or simply a module when the ring is understood.

If M and N are R-modules, then a mapping $\sigma : M \to N$ is an *R-morphism* or a *morphism* if $\sigma(rm + sn) = r\sigma(m) + s\sigma(n)$ for r and s in R and m and n in M. The *kernel* of σ, $\mathrm{Ker}(\sigma)$, is the set $\{m \text{ in } M \mid \sigma(m) = 0\}$, and the *image* of σ, $\mathrm{Im}(\sigma) = \sigma(M)$ is the set $\{\sigma(m) \mid m \text{ in } M\}$. The morphism σ is *surjective* if $\mathrm{Im}(\sigma) = N$, *injective* if $\mathrm{Ker}(\sigma) = 0$, and *bijective* if it is both injective and surjective. If $\sigma : M \to N$ is a bijective R-morphism, we say M is *isomorphic* to N, denoted $M \simeq N$.

Let M be an R-module. Let N be a subset of M. Then N is a *submodule* of M if N is a subgroup of M and for each r in R, $rN = \{rn \mid n \in N\}$ is a subset of N. If N is a submodule of M then the quotient group M/N is a module under $r(m + N) = rm + N$ and is called the quotient module. Observe there is a natural surjective R-morphism $\sigma : M \to M/N$ given by $m \to m + N$.

Let M be an R-module and N_α , $\alpha \in \Lambda$, a collection of submodules of M . Then

$$\bigcap_{\alpha\in\Lambda} N_\alpha = \{n \mid n \in N_\alpha \text{ for all } \alpha \in \Lambda\}$$

and

$$\sum_{\alpha\in\Lambda} N_\alpha = \{\sum_f r_\alpha n_\alpha \mid r_\alpha \in R , n_\alpha \in N_\alpha\}$$

are submodules of M called the intersection and sum ($\sum_f$ indicates the sums are finite), respectively.

Let T be a subset of an R-module M . Let

$$RT = \{\sum_f r_\alpha t_\alpha \mid r_\alpha \in R , t_\alpha \in T\} .$$

Then RT is a submodule of M called the module generated by T . We sometimes write RT as $\langle T\rangle$. If $RT = M$ we say M is generated by T . If the cardinality $|T|$ of T is finite, i.e., $|T| < \infty$, and $RT = M$, then M is said to be finitely generated. If $T = \langle t\rangle$ and $RT = Rt = \{rt \mid r \in R\}$ is equal to M we say M is a cyclic module.

If $\sigma : M \to M$ is an R-morphism then σ is called an endomorphism of M . If σ is bijective then σ is an automorphism of M . If

$\sigma^n = \sigma \cdots \sigma$ (n-factors) is equal to 0 then σ is called nilpotent.
Let

$$\mathrm{End}_R(M) = \{\sigma : M \to M \mid \sigma \text{ is an R-morphism}\}$$

and

$$\mathrm{GL}_R(M) = \{\sigma \in \mathrm{End}_R(M) \mid \sigma \text{ is an automorphism}\} .$$

Note that $\mathrm{End}_R(M)$ is a ring and $\mathrm{GL}_R(M)$ is a group called the R-endomorphism ring of M and the general linear group of M, respectively.

Let

$$\mathrm{Hom}_R(M,N)$$

denote the set of R-morphisms $\sigma : M \to N$ for modules M and N.

If $\sigma : M \to N$ is a surjective R-morphism then $N \simeq M/\mathrm{Ker}(\sigma)$ and there is a natural bijective order preserving correspondence between the submodules of N and the submodules of M containing $\mathrm{Ker}(\sigma)$. We refer to this fact as the Correspondence Theorem. Further, if N and P are submodules of M then

(a) if $P \subseteq N \subseteq M$, then $M/N \simeq (M/P)/(N/P)$.

(b) $(P + N)/P \simeq N/(P \cap N)$.

Let M be an R-module with submodules $N_1,\ldots,N_t$. Call M a direct sum

(<u>internal</u>) of $N_1,\ldots,N_t$ if every element m of M can be expressed uniquely as

$$m = n_1 + \cdots + n_t$$

with n_i in N_i . Equivalently, $M = \sum_i N_i$ and $N_i \cap (\sum_{j\neq i} N_j) = 0$ for $1 \leq i \leq t$.

On the other hand, if $N_1,\ldots,N_t$ are R-modules, then the set of t-tuples $\langle n_1,\ldots,n_t\rangle$ with n_i in N_i is an R-module with component-wise addition and $r\langle n_1,\ldots,n_t\rangle = \langle rn_1,\ldots,rn_t\rangle$. This module is the <u>direct sum</u> (<u>external</u>) of $N_1,\ldots,N_t$.

If M is a direct sum of $N_1,\ldots,N_t$ then M is written as $M = \oplus \sum_{i=1}^{t} N_t = N_1 \oplus \cdots \oplus N_t$.

Let $\{M_i\}$ be a collection of R-modules and $\{\sigma_i\}$ a collection of R-morphisms with $\sigma_i : M_i \to M_{i+1}$. Then the sequence

$$\cdots \longrightarrow M_{i-1} \xrightarrow{\sigma_{i-1}} M_i \xrightarrow{\sigma_i} M_{i+1} \longrightarrow \cdots$$

is called <u>exact</u> if for each i , $\mathrm{Im}(\sigma_{i-1}) = \mathrm{Ker}(\sigma_i)$. An exact sequence $M \xrightarrow{\sigma} N \longrightarrow 0$ $(0 \longrightarrow N \xrightarrow{\sigma} M)$ is said to <u>split</u> or <u>splits</u> if there exists an R-morphism $\beta : N \to M$ $(\beta : M \to N)$ with $\sigma\beta =$ identity on N ($\beta\sigma =$ identity on N). If $M \xrightarrow{\sigma} N \longrightarrow 0$ splits with splitting morphism $\beta : N \to M$ then $N \simeq \mathrm{Im}(\beta)$, $\mathrm{Im}(\beta) \cap \mathrm{Ker}(\sigma) = 0$, and $M = \mathrm{Im}(\beta) \oplus \mathrm{Ker}(\sigma)$. If $0 \longrightarrow N \xrightarrow{\sigma} M$ splits then $M = \mathrm{Im}(\sigma) \oplus \mathrm{Ker}(\beta)$.

(B) LOCAL RINGS AND THEIR SPACES

Let R be a commutative ring and $\text{Rad}(R) = \cap m$ where the intersection extends over all maximal ideals m of R.

A commutative ring R is said to be <u>local</u> if R satisfies any of the following equivalent conditions:

(a) $R/\text{Rad}(R)$ is a field,

(b) R has only one maximal ideal,

(c) all non-units of R are contained in a single proper ideal,

(d) the non-units of R form a proper ideal,

(e) for every r in R either r or $1 - r$ is a unit.

Some examples of local rings:

(a) Any field is a local ring.

(b) Z/Zp^n is a local ring where Z denotes the rational integers and p is a prime.

(c) $k[X]/(f^n)$ where k is a field and f an irreducible polynomial.

(d) Any Artinian commutative ring (i.e., descending chain condition on ideals) is a direct product of local rings.

(e) If R is a local ring and A an ideal of R, then R/A is local.

(f) If R is a commutative ring and P is a prime ideal, then the ring of fractions $S^{-1}R$ of R localized at the complement

$S = R - P$ of P is local.

(g) Let R be a commutative ring and m a maximal ideal of R. Then R/m^t is local.

Let R be a local ring. Let V be a finitely generated R-module. Then V is an R-space of finite dimension (or R-free) if there exist $b_1,\ldots,b_n$ in V with

(a) $V = Rb_1 \oplus \cdots \oplus Rb_n$

(b) the map $R \to Rb_i$ by $1 \to b_i$ is an R-isomorphism for $1 \le i \le n$.

The set $\{b_1,\ldots,b_n\}$ is called an R-basis for V. If $V = Rb_1 \oplus \cdots \oplus Rb_n$ is an R-space then there exists a natural R-isomorphism $V \to R^{(n)} = R \oplus \cdots \oplus R$ (n-summands) by $\sum r_i b_i \to \langle r_1,\ldots,r_n\rangle$.

If A is an ideal of R, there is a natural R-morphism $R^{(n)} \to (R/A)^{(n)}$ given by $\langle r_1,\ldots,r_n\rangle \to \langle r_1 + A,\ldots,r_n + A\rangle$. Further for the R-space $V = \oplus \sum Rb_i$, the diagram

$$\begin{array}{ccc} V & \longrightarrow & R^{(n)} \\ \downarrow & & \downarrow \\ V/AV & \longrightarrow & (R/A)^{(n)} \end{array}$$

is commutative.

If $V = Rb_1 \oplus \cdots \oplus Rb_n$ is an R-space and M is any R-module, then any map of $B = \{b_1,\ldots,b_n\} \to M$ extends to a unique R-morphism $V \to M$ as follows: if $b_i \to m_i$ then $\sum r_i b_i \to \sum r_i m_i$. This gives a natural bijection between Map(B,M) and $\mathrm{Hom}_R(V,M)$.

<u>(I.1)</u> <u>THEOREM</u>. (Nakayama's Lemma) Let M be a finitely generated module over a local ring R . Let m denote the maximal ideal of R . Then

(a) if A is a proper ideal of R and AM = M , then M = 0 .

(b) if A is a proper ideal of R and N is a submodule of M with M = AM + N , then M = N .

<u>Proof</u>. (a) Suppose M = AM and $u_1, \ldots, u_n$ is a generating set of M for which n is minimal. Then u_1 is in M = AM . Hence

$$u_1 = a_1u_1 + \cdots + a_nu_n \qquad (a_i \text{ in } A) .$$

Thus $(1 - a_1)u_1 = a_2u_2 + \cdots + a_nu_n$. Since a_1 is in $A \subseteq m$, $1 - a_1$ is a unit. Thus u_1 is in $Ru_2 + \cdots + Ru_n$, contradicting the minimality of n . For (b), apply (a) to M/N and note A(M/N) = (AM + N)/N = M/N .

<u>(I.2)</u> <u>THEOREM</u>. Let R be a local ring with maximal ideal m . Let M be a finitely generated R-module.

(a) A subset $\{u_i\}_{i=1}^n$ of M is a generating set for M if and only if their residue classes $\{\bar{u}_i\}_{i=1}^n$ generate the R/m-vector space M/mM .

(b) A subset $\{u_i\}_{i=1}^n$ of M is a minimal generating set for M if and only if $\{\bar{u}_i\}_{i=1}^n$ form an R/m-basis for M/mM .

(c) Any generating set of M contains a minimal generating set. Further, if $\{u_1,\ldots,u_n\}$ and $\{v_1,\ldots,v_m\}$ are both minimal generating sets for M , then $m = n$ and there is an R-automorphism $\sigma : R^{(n)} \to R^{(n)}$ with $\sigma \langle u_1,\ldots,u_n\rangle = \langle v_1,\ldots,v_n\rangle$.

Proof. Simply set $N = \sum Ru_i$ and use Nakayama's Lemma.

(I.3) COROLLARY. Let R be local and V be an R-space having bases $\{b_1,\ldots,b_n\}$ and $\{c_1,\ldots,c_m\}$. Then $n = m$ and we say V has dimension n , $\dim(V) = n$.

(I.4) COROLLARY. Let R be local and V be an R-space of finite dimension. Let W be a direct summand of V . Then W is an R-space.

Proof. Let $\{u_1,\ldots,u_n\}$ be a minimal set of generators of W . Let F be an R-space with basis $\{b_1,\ldots,b_n\}$ and define $\sigma : F \to W$ by $\sigma b_i = u_i$. We have an exact sequence

$$0 \longrightarrow U \longrightarrow F \xrightarrow{\sigma} W \longrightarrow 0$$

where $U = \mathrm{Ker}(\sigma)$. Suppose $\sum r_i b_i$ is in U and not in mF where m is the maximal ideal of R . Then some coefficient r_i is a unit. Without loss, assume r_1 is a unit. But $\sigma(\sum r_i b_i) = \sum r_i u_i = 0$. Hence, solve for u_1 in terms of $u_2,\ldots,u_n$.

This contradicts minimality of n . Thus $U \subset mF$. The module W is a direct summand of V . Thus,

$$\begin{array}{ccccc} & & V & & \\ & & \downarrow \Pi_W = \text{proj. on } W & & \\ & & W & & \\ & & \downarrow \text{id} = \text{identity} & & \\ F & \xrightarrow{\sigma} & W & \longrightarrow & 0 \end{array}$$

Since V is free there is a natural morphism $\beta^* : V \to F$ with $\sigma\beta^* = \text{id}\ \Pi_W$. Take $\beta = \beta^*|_W$. Thus $\beta : W \to F$ and $\sigma\beta = \text{id}$, i.e., $F \xrightarrow{\sigma} W \longrightarrow 0$ splits. Hence $F = W^* \oplus \text{Ker}(\sigma)$ where $W^* = \text{Im}(\beta) \simeq W$. Then $mF = mW^* \oplus m\,\text{Ker}(\sigma)$. But $mU \subset U \subset mF$ where $U = \text{Ker}(\sigma)$. Thus $U = mU \oplus (U \cap mW^*)$. Since $U \cap mW^* \subset U \cap W^* = 0$, we have $U = mU$. By Nakayama's Lemma, $U = 0$ and $F \simeq W$.

Let V be a space over a local ring R of finite dimension. A direct summand W of V is an R-*subspace* of V . By the above result subspaces are themselves spaces.

(I.5) COROLLARY. Let V be a finite dimensional space over a local ring R . Then a basis of a subspace may be extended to a basis of the space.

Let R be local and V be an R-space of dimension n . A subspace of V

(a) of dimension 1 is a *line*.

(b) of dimension 2 is a *plane*.

(c) of dimension $n - 1$ is a hyperplane.

The next result is straightforward and characterizes generators of lines.

(I.6) THEOREM. (On unimodular elements) Let R be a local ring with maximal ideal m and $k = R/m$. Let V be an R-space and x be in V . The following are equivalent.

(a) If $x = \sum r_i b_i$ for a basis $\{b_1,\ldots,b_n\}$ of V , then $(r_1,\ldots,r_n) = R$.

(b) $L = Rx$ is a line.

(c) There is a morphism $\sigma : V \to R$ with $\sigma x = 1$.

(d) The mapping $R \to V$ by $1 \to x$ gives a split exact sequence $0 \to R \to V$.

(e) Under $\Pi : V \to V/mV$, we have $\Pi x \neq 0$.

An element x of V satisfying the above theorem is called a unimodular element of V .

We look at some examples.

(a) An element in an R-space may be R-free but not unimodular, i.e., there may be x in V with $Rx \simeq R$ but Rx not a summand of V . Take $Z_p = \{\frac{a}{b} \mid (b,p) = 1\}$ where a and b are integers and p is a prime. Take $V = Z_p \oplus Z_p = \{\langle a,b\rangle \mid a,b \text{ are in } Z_p\}$. Let $x = \langle p,0\rangle$. Then x is Z_p-free since Z_p is a domain. But under

$\Pi : V \to V/(p)V$, $\Pi x = 0$.

(b) An element of a space V may not be contained in any line. Let $R = (Z/Zp)[X,Y]/(X^2,XY,Y^2) = \{a + bX + cY \mid a,b,c \text{ in } Z/Zp\}$. Let $V = R \oplus R$ and take $x = \langle X,Y\rangle$. Suppose there is a b in V with b unimodular and $rb = x$ for some r in R . Then $b = \langle s_1,s_2\rangle$ and $rs_1 = X$, $rs_2 = Y$. Since b is unimodular either s_1 or s_2 is a unit. If s_1 is a unit, then $r = s_1^{-1}X$ and $s_1^{-1}s_2X = Y$, but no multiple of X in R gives Y . A similar argument suffices if s_2 is a unit.

(c) Two lines may intersect in more than one point. Take $R = Z/Zp^n$, $V = R \oplus R$, $e = \langle 1,0\rangle$ and $f = \langle 1,p^{n-1}\rangle$. Then Re and Rf are distinct lines, but $pe = pf$, $p^2e = p^2f,\ldots,p^ne = p^nf$.

Let R be a local ring with maximal ideal m . The quotient field $k = R/m$ is called the <u>residue field of</u> R . Let V be a finite dimensional R-space. Then the quotient module V/mV is naturally a k-vector space. We denote the natural map $V \to V/mV$ by Π .

We examine hyperplanes.

<u>(I.7)</u> <u>PROPOSITION</u>. Let V be a space of dimension n over a local ring R . Let U and W be hyperplanes. If $\Pi U \neq \Pi W$ under $\Pi : V \to V/mV$ then $U \cap W$ is a subspace of dimension $n - 2$.

<u>Proof</u>. Let $V = Ra \oplus W = Rb \oplus U$. Let $W = \oplus \sum_{i=2}^{n} Ra_i$ and $U = \oplus \sum_{i=2}^{n} Rb_i$. It is easy to see, since $\Pi U \neq \Pi W$, that $V = U + W$. Since $\Pi U \neq \Pi W$, assume Πa_2 is not in ΠU . It is straightforward to show $\Pi W = k\Pi a_2 \oplus (\Pi U \cap \Pi W)$. We claim $W = Ra_2 \oplus (W \cap U)$. Let $a_2 = \alpha b + \beta$ where α is in R and not in m and β is in U . Then β is in $Ra_2 \oplus U$. Thus $V = Ra_2 \oplus U$. Let x be arbitrary in W . Then $x = \delta a_2 + c$ where δ is in R and c is in U . Thus c is in $U \cap W$. Then x is in $Ra_2 + (U \cap W)$. Hence $W = Ra_2 + (U \cap W)$. Also $Ra_2 \cap (U \cap W) \subseteq U \cap Ra_2 = 0$. Thus $W = Ra_2 \oplus (W \cap U)$.

<u>(C)</u> <u>LINEAR MAPS OF R-SPACES</u>

Throughout this section let R denote a local ring with maximal ideal m and residue field $k = R/m$. Let V be an R-space of dimension n . Let V have a basis $b_1,\ldots,b_n$.

Let $\sigma : V \to V$ be an R-morphism. The map

$$\mathrm{Mat} : \mathrm{End}_R(V) \to (R)_n$$

from the endomorphism ring of V to the ring of n by n matrices over R given by

$$\mathrm{Mat} : \sigma \to \mathrm{Mat}(\sigma) = [a_{ji}]$$

where

$$\sigma b_i = \sum_{j=1}^{n} a_{ji} b_j \qquad 1 \le i \le n$$

gives a natural R-algebra isomorphism. Further, if $x = \sum a_i b_i$ is identified with $R^{(n)}$ under $\sum a_i b_i \to \langle a_1, \ldots, a_n \rangle^t$ (t denotes the transpose), then $\sigma x \to \mathrm{Mat}(\sigma)\langle a_1, \ldots, a_n \rangle^t$. If A is an ideal of R, then

$$\begin{aligned} \mathrm{End}_R(V/AV) &\simeq \mathrm{End}_{R/A}((R/A)^{(n)}) \\ &\simeq \mathrm{End}_R(V)/(A) \\ &\simeq (R)_n/(A)_n \end{aligned}$$

where $(A) = \{\sigma \text{ in } \mathrm{End}(V) \mid \sigma V \subseteq AV\}$ and $(A)_n$ denotes the ideal in $(R)_n$ of n by n matrices over A. The map

$$A \to (A) \qquad (\text{resp., } A \to (A)_n)$$

gives a lattice isomorphism between the ideals of R and the ideals of End(V) (resp., ideals of $(R)_n$ — see (I.F)).

The determinant is a multiplicative map $\mathrm{End}(V) \to R$ satisfying: σ is in GL(V) if and only if $\det(\sigma)$ is a unit of R. Indeed, if R^* denotes the units of R and $SL_R(V) = \mathrm{Ker}(\det)$ then det gives an exact sequence

$$1 \longrightarrow SL(V) \xrightarrow{\text{id}} GL(V) \xrightarrow{\det} R^* \longrightarrow 1 .$$

The group SL(V) is the <u>special linear group</u> of V .

The center(End(V) $\simeq R$ under $rI \to r$.

Let $V^* = \mathrm{Hom}_R(V,R)$ denote the <u>dual space</u> of V. If σ is in $\mathrm{End}(V)$ then there is a natural R-morphism σ^t in $\mathrm{End}(V^*)$ given by $\sigma^t(f) = f\sigma$. If V has a basis $\{b_1,\ldots,b_n\}$ then a <u>dual basis</u> for V^* is constructed by defining $\{b_1^*,\ldots,b_n^*\}$ by $b_i^*(b_j) = \delta_{ij}$. If σ has matrix $\mathrm{Mat}(\sigma) = [a_{ji}]$ relative to $\{b_1,\ldots,b_n\}$ then σ^t has matrix $[b_{ji}]$ where $b_{ji} = a_{ij}$ relative to the basis $\{b_1^*,\ldots,b_n^*\}$. Denote $[b_{ji}]$ by $[a_{ji}]^t$ and call this the <u>transpose</u> of $[a_{ji}]$. Finally, $V^{**} = (V^*)^*$ is identified with V under $\langle b_i^*, b_j\rangle = b_i^*(b_j)$.

We examine the generators of $\mathrm{End}(V)$ and $GL(V)$. Let $\{b_1,\ldots,b_n\}$ be a basis of V. Define

(a) $\tau_{ji}(\lambda) : V \to V \quad (j \neq i)$ (<u>elementary transvection</u>)

by

$$\tau_{ji}(\lambda) : \begin{cases} b_i \to b_i + \lambda b_j & \\ b_k \to b_k & k \neq i, \end{cases}$$

(b) $d_i(\lambda) : V \to V$ (<u>elementary dilation</u>)

by

$$d_i(\lambda) : \begin{cases} b_i \to \lambda b_i & (\lambda \text{ a unit of } R) \\ b_j \to b_j & j \neq i, \end{cases}$$

(c) $E_{ij} : V \to V$ (<u>elementary matrix unit</u>)

by

$$E_{ij} : b_k \to \delta_{jk} b_i \qquad (\delta_{jk} = \text{Kronecker delta}),$$

(d) $\rho_{ij} : V \to V$ (<u>elementary permutation</u>)

by

$$\rho_{ij} : \begin{cases} b_i \to b_j & \\ b_j \to b_i & \\ b_k \to b_k & k \neq i,j \ . \end{cases}$$

Clearly all of the above are R-morphisms of V and $\tau_{ji}(\lambda)$, $d_i(\lambda)$ and ρ_{ij} are in $GL(V)$. Further,

(a) $d_i(\lambda)^{-1} = d_i(\lambda^{-1})$, $d_i(\lambda)d_i(\beta) = d_i(\lambda\beta)$.

(b) $\tau_{ji}(\lambda)^{-1} = \tau_{ji}(-\lambda)$, $\tau_{ji}(\lambda)\tau_{ji}(\beta) = \tau_{ji}(\lambda + \beta)$.

(c) $E_{ji}E_{qp} = \delta_{iq}E_{jp}$, $\sum E_{ii} = I$.

(d) $\rho_{ij}\rho_{ij} = (\delta_{ij}) = \sum_{i,j} \delta_{ij}E_{ij}$.

If $\sigma : V \to V$ satisfies

$$\sigma b_i = \sum_j a_{ji}b_j \qquad 1 \leq i \leq n ,$$

then

$$\sigma = \sum_{j,i} a_{ji}E_{ji} .$$

Indeed, $\{E_{ji}\}$ form a free R-basis of $End(V)$. With the aid of the elementary matrix units we can classify the ring automorphisms of $End(V)$.

A set $\{F_{ij}\}$, $1 \leq i,j \leq n$, of elements of $End(V)$ is a set of <u>matrix units</u> if

(a) $F_{ij} = F_{\ell m}$ only if $i = \ell$ and $j = m$,

(b) $F_{ij}F_{\ell m} = \delta_{j\ell}F_{im}$ ($\delta_{j\ell}$ = Kronecker delta) ,

(c) $\sum_i F_{ii} = I$.

If $\{F_{ij}\}$ is a set of matrix units and ρ is in GL(V) then $\{\rho F_{ij}\rho^{-1}\}$ is a set of matrix units. We show the converse of this statement.

<u>(I.8)</u> <u>THEOREM</u>. If $\{F_{ij}\}$ and $\{G_{ij}\}$ are two sets of matrix units in End(V) , then there is a ρ in GL(V) with $\rho F_{ij}\rho^{-1} = G_{ij}$ for $1 \leq i,j \leq n$. That is, GL(V) acts transitively under conjugation on the sets of matrix units in End(V) .

<u>Proof</u>. Let $\{b_1,\ldots,b_n\}$ be a basis of V and $\{E_{ij}\}$ denote the elementary matrix units relative to this basis. Let $\{F_{ij}\}$ be an arbitrary set of matrix units. Let v be in V . Then $v = Iv = \sum_i F_{ii}v = \sum_i F_{i1}F_{1i}v = \sum_i F_{i1}v_i$ where $v_i = F_{1i}v$. Thus $V = \sum_i F_{i1}V$. Since $F_{ij}F_{k\ell} = \delta_{jk}F_{i\ell}$, it is also easy to see that $V = F_{11}V \oplus \cdots \oplus F_{n1}V$. But each $F_{i1}V$ is a direct summand of V , hence a subspace. Thus $F_{i1}V$ is free of dimension one — a line. Let $L_i = F_{i1}V$. The matrix units $F_{ji}|_{L_i} : L_i \to L_j$ and $F_{ij}|_{L_j} : L_j \to L_i$ and have as their composition the identity in both directions. Thus $F_{ji}|_{L_i}$ is an R-isomorphism. Then if $L_1 = Rb$, b a basis of L_1 , we have $F_{j1}b$ a basis element of L_j . Setting $c_1 = b$ and $c_j = F_{j1}b$ for $j \geq 2$, we have a basis $\{c_1,\ldots,c_n\}$ of V . Define σ in GL(V) by $\sigma : b_i \to c_i$ (change of basis transformation). Then

$$\begin{aligned}(\sigma^{-1}F_{ij}\sigma)b_k &= (\sigma^{-1}F_{ij})c_k \\ &= \sigma^{-1}F_{ij}F_{k1}b \\ &= \sigma^{-1}\delta_{jk}F_{i1}b \\ &= \delta_{jk}\sigma^{-1}c_i \\ &= \delta_{jk}b_i = E_{ij}b_k .\end{aligned}$$

Thus $\sigma^{-1}F_{ij}\sigma = E_{ij}$.

Theorem (I.8) allows the determination of the automorphisms of End(V) .

Let $\alpha : R \to R$ be a ring morphism. Let $\sigma : V \to V$ be an R-morphism where $\sigma b_i = \sum_j a_{ji}b_j$. Define σ^α by

$$(\sigma^\alpha)(b_i) = \sum_j \alpha(a_{ji})b_j .$$

<u>(I.9)</u> <u>THEOREM</u>. Let $\Lambda : \text{End}(V) \to \text{End}(V)$ be a ring isomorphism. Then there is a ring isomorphism $\alpha : R \to R$ and a β in GL(V) such that

$$\Lambda\sigma = \beta^{-1}\sigma^\alpha\beta$$

for all σ in End(V) .

<u>Proof</u>. Let σ be in End(V) and $\{E_{ij}\}$ a set of elementary matrix units. Thus $\sigma = \sum a_{ij}E_{ij}$. Therefore, $\Lambda\sigma = \Lambda(\sum a_{ij}E_{ij}) = \Lambda(\sum (a_{ij}I)E_{ij}) = \sum (\Lambda a_{ij}I)(\Lambda E_{ij})$. Let $\alpha = \Lambda|_{\text{center }(\text{End}(V))}$. Then $\alpha : R \to R$ is a ring isomorphism. Further, $\Lambda E_{ij} = F_{ij}$ is a set of

matrix units. Thus, by (I.8) there is a β in $GL(V)$ satisfying $\beta F_{ij}\beta^{-1} = E_{ij}$.

Thus

$$\begin{aligned}\Lambda\sigma &= \sum \Lambda(a_{ij}I)\Lambda E_{ij} \\ &= \sum (\alpha a_{ij})F_{ij} \\ &= \sum (\alpha a_{ij})\beta^{-1}E_{ij}\beta \\ &= \beta^{-1}(\sum \alpha a_{ij}E_{ij})\beta \\ &= \beta^{-1}\sigma^{\alpha}\beta \qquad .\end{aligned}$$

With the aid of the elementary transvections and dilations we generate $GL(V)$.

<u>(I.10)</u> <u>THEOREM</u>. The group $GL(V)$ is generated by

$$\{\tau_{ij}(\lambda), d_k(\mu) \mid \lambda \text{ in } R , \mu \text{ in } R^* , i \neq j\} .$$

(Assume $n \geq 2$.)

<u>Proof</u>. Observe that

$$\begin{bmatrix} 1 & 0 & \cdots & & 0 \\ b_2 & 1 & \ddots & & \vdots \\ \vdots & \vdots & & \ddots & \\ & & & 1 & 0 \\ b_n & 0 & \cdots & 0 & 1 \end{bmatrix} = \tau_{21}(b_2)\cdots\tau_{n1}(b_n)I .$$

For simplicity identify $GL(V)$ with $GL_n(R)$ the group of n by n invertible matrices in $(R)_n$ relative to a fixed (but arbitrary) basis. Let $A = [a_{ij}]$ be in $GL_n(R)$. Then $\det(A)$ is a unit. But $\det(A) = \sum_j a_{1j}A_{1j}$ where A_{1j} is a cofactor (expansion of the determinant by minors).

Thus one of the a_{1j} is a unit (since the sum of non-units in a local ring cannot be a unit). Since R is local it is now easy to find $b_2,\ldots,b_n$ in R with $a_{11} + b_2a_{12} + \cdots + b_na_{1n} = u$ where u is a unit. That is,

$$A \begin{bmatrix} 1 & 0 & \cdots & 0 \\ b_2 & 1 & & \vdots \\ \vdots & & \ddots & \vdots \\ b_n & 0 & \cdots & 1 \end{bmatrix} = \begin{bmatrix} u & * & & * \\ * & * & & \\ & & & \\ * & & & * \end{bmatrix} .$$

Without loss, assume that a_{11} is a unit. If $A = [a_{ij}]$ is multiplied on the left by

$$\tau_{21}(a_{11}^{-1}(1 - a_{21}))$$

and then by

$$\tau_{12}(1 - a_{11}) ,$$

obtain a = 1 in the (1,1)-position. By left multiplication by suitable elementary transvections, sweep out the first column of A leaving

$$\begin{bmatrix} 1 & * & \cdots & * \\ 0 & * & \cdots & * \\ \vdots & \vdots & & \vdots \\ 0 & * & \cdots & * \end{bmatrix} .$$

Continuing this process, eventually one arrives at the matrix

$$\begin{bmatrix} 1 & 0 & \cdots & 0 & a'_{1n} \\ 0 & 1 & \cdots & 0 & a'_{2n} \\ \vdots & \vdots & & & \vdots \\ 0 & 0 & \cdots & & w \end{bmatrix} .$$

Left multiply by $\tau_{in}(w^{-1}(-a'_{in}))$, $1 \leq i \leq n - 1$, and place 0's in the last column except for w in the (n,n)-position. Thus by elementary transvections

$$\tau_1 \cdots \tau_t A = \begin{bmatrix} 1 & 0 & \cdots & 0 \\ 0 & 1 & & \cdot \\ \vdots & & & \cdot \\ & & & \cdot \\ 0 & \cdot & \cdot \; \cdot & w \end{bmatrix} = d_n(w) .$$

Since $\tau_{ij}(\lambda)^{-1} = \tau_{ij}(-\lambda)$, $A = \tau_t^{-1} \cdots \tau_1^{-1} d_n(w)$ is a product of transvections and a dilation.

Observe $w = \det(A)$ since $\det(\tau_{ij}(\lambda)) = 1$.

(I.11) COROLLARY.

(a) Let $T(V)$ denote the group generated by the elementary transvections (relative to a given basis). Then

$$T(V) = SL(V) .$$

Thus $T(V)$ does not depend on the basis.

(b) Let R^* denote $\{d_n(w) \mid w$ in the units of $R\}$. Then

$$GL(V) = T(V)R^* = SL(V)R^* .$$

(c) Let A be an ideal of R . Then the natural group morphism $GL(V) \to GL(V/AV)$ is surjective since $T(V) \to T(V/AV)$ is obviously surjective. This gives a natural commutative diagram

$$\begin{array}{ccccccccc} 1 & \longrightarrow & SL(V) & \longrightarrow & GL(V) & \longrightarrow & R^* & \longrightarrow & 1 \\ & & \downarrow \text{surj.} & & \downarrow \text{surj.} & & \downarrow \text{surj.} & & \\ 1 & \longrightarrow & SL(V/AV) & \longrightarrow & GL(V/AV) & \longrightarrow & (R/A)^* & \longrightarrow & 1 \end{array}$$

with exact rows.

(D) SEMI-LINEAR MAPS AND THE FUNDAMENTAL THEOREM OF PROJECTIVE GEOMETRY

Let V be a free R-module, i.e., an R-space, of dimension n over a local ring R . A <u>semi-linear</u> morphism $\phi : V \to V$ with <u>associated ring automorphism</u> $\sigma : R \to R$ is a mapping satisfying

(a) $\phi(v + w) = \phi(v) + \phi(w)$ for all v and w in V ,

(b) $\phi(rv) = \sigma(r)\phi(v)$ for r in R and v in V .

If $\sigma = id_R$, then ϕ is a linear map.

Let ϕ be σ-semi-linear and $\{b_1,\ldots,b_n\}$ be a basis for V . Then

$$\phi(b_i) = \textstyle\sum_j a_{ji}b_j \ .$$

If $A = Mat(\phi) = [a_{ij}]$, then identify ϕ with the pair $\langle A,\sigma\rangle$,

$$\phi \longleftrightarrow \langle A,\sigma\rangle \ .$$

If $\sum r_i b_i$ is an element of V , then

$$\begin{aligned}\phi(\textstyle\sum r_i b_i) &= \textstyle\sum \sigma(r_i)\phi(b_i)\\ &= \textstyle\sum_i \sigma(r_i)(\sum_j a_{ji}b_j)\\ &= \textstyle\sum_j (\sum_i a_{ji}\sigma(r_i))b_j \ .\end{aligned}$$

Thus, in terms of matrices,

$$\langle A,\sigma\rangle \begin{bmatrix} r_1 \\ r_2 \\ \vdots \\ r_n \end{bmatrix} = A \begin{bmatrix} \sigma r_1 \\ \sigma r_2 \\ \vdots \\ \sigma r_n \end{bmatrix} \ .$$

Further, ϕ is invertible if and only if A is invertible. If ϕ and ρ are two semi-linear morphisms with

$$\phi \longleftrightarrow \langle A,\sigma\rangle$$

and

$$\rho \longleftrightarrow \langle B,\tau\rangle \ ,$$

then their product

$$\phi\rho \longleftrightarrow \langle A,\sigma\rangle\langle B,\tau\rangle$$
$$= \langle A(B^{\sigma}),\sigma\tau\rangle$$

where if $B = [b_{ij}]$ then $B^{\sigma} = [\sigma b_{ij}]$.

If $\{c_1,\ldots,c_n\}$ is another basis for V with $c_i = \sum_j d_{ji}b_j$, then the "matrix" for ϕ relative to $\{c_1,\ldots,c_n\}$ is given by

$$\langle D,i\rangle^{-1}\langle A,\sigma\rangle\langle D,i\rangle = \langle D^{-1}AD^{\sigma},\sigma\rangle$$

where $D = [d_{ij}]$.

Utilizing semi-linear morphisms of V , we may restate (I.9).

<u>(I.12)</u> <u>THEOREM</u>. Let $\wedge : \mathrm{End}(V) \to \mathrm{End}(V)$ be a ring isomorphism. Then there is a semi-linear isomorphism $\phi : V \to V$ such that

$$\wedge\sigma = \phi^{-1}\sigma\phi$$

for all σ in $\mathrm{End}(V)$.

Let

$$\mathrm{SEnd}(V) = \{\phi : V \to V \mid \phi \text{ is semi-linear}\}$$

and

$$\mathrm{SGL}(V) = \{\phi \text{ in } \mathrm{SEnd}(V) \mid \phi \text{ is invertible}\} .$$

Observe there is a natural exact sequence of groups

$$1 \longrightarrow GL(V) \xrightarrow{\text{inj}} SGL(V) \xrightarrow{\lambda} \mathrm{Aut}(R) \longrightarrow 1$$

where $\text{inj} : A \to \langle A,1\rangle$ and $\lambda : \langle A,\sigma\rangle \to \sigma$. We examine this sequence in greater detail in the next section.

Semi-linear mappings arise naturally in the study of projectivities or collineations of the projective space of V.

Let $P(V)$ denote the set of lines of V, i.e., the set of all summands of V of dimension 1 (equivalently, the set of all Re for e unimodular in V). $P(V)$ is called the <u>projective space</u> of V. A mapping $\alpha : P(V) \to P(V)$ is a <u>projectivity</u> or a <u>collineation</u> if

(a) α is bijective.

(b) For lines L, L_1 and L_2:

  (1) $L_1 + L_2$ is a plane if and only if $\alpha L_1 + \alpha L_2$ is a plane.

  (2) Suppose $L_1 + L_2$ is a plane. Then $L \subset L_1 + L_2$ if and only if $\alpha L \subset \alpha L_1 + \alpha L_2$.

Thus, a projectivity is a bijection between lines which "preserves" planes.

If ϕ is an invertible semi-linear morphism of V then $\phi(e)$ is unimodular if and only if e is unimodular. This implies that ϕ induces naturally a projectivity $P(\phi)$ on $P(V)$ by

$$P(\phi)Re = R\phi(e)$$

where e is unimodular.

The Fundamental Theorem of Projective Geometry states that conversely every projectivity is induced by a semi-linear isomorphism of V.

(I.13) THEOREM. (Fundamental Theorem of Projective Geometry)[1] Let V be an R-space of dimension ≥ 3. Let $\alpha : P(V) \to P(V)$ be a projectivity. Then, there is a semi-linear isomorphism $\phi : V \to V$ satisfying

$$P(\phi) = \alpha .$$

Proof. The proof will be provided in a series of steps.

(a) Let $\{b_1,\ldots,b_n\}$ be a basis for V. Suppose $\alpha Rb_i = Rc_i$. We claim that $\{c_1,\ldots,c_n\}$ is a basis for V.

Let $Rb \subset Rb_1 + \cdots + Rb_i$ be a line. By induction, we show $\alpha Rb \subset Rc_1 + \cdots + Rc_i$. If $i = 1$ the claim is trivial and if $i = 2$ the statement follows from the definition of a projectivity. Suppose $b = \alpha_1 b_1 + \cdots + \alpha_r b_r$. Since b is unimodular some α_i, say α_1, is a unit. Then $Rb \subseteq R(b - \alpha_r b_r) + Rb_r$. Thus

[1] For a more general statement of this theorem see (II.22).

$\alpha Rb \subseteq \alpha R(b - \alpha_r b_r) + \alpha Rb_r$. By induction $\alpha R(b - \alpha_r b_r) \subseteq Rc_1 + \cdots + Rc_{r-1}$. Hence $\alpha Rb \subseteq Rc_1 + \cdots + Rc_r$. Thus if $L \subset V$ then $\alpha L \subset Rc_1 + \cdots + Rc_n$. Since α is surjective, $\{c_1, \ldots, c_n\}$ spans V and by (I.2) forms a basis for V .

(b) Now select the images of $Rb_1, \ldots, Rb_n$ under α more carefully.

Let $2 \leq i \leq n$. Then $\alpha R(b_1 + b_i) = R(\beta c_1 + \delta c_i)$ for some β and δ in R . Since $Rb_1 \subseteq R(b_1 + b_i) + Rb_i$ we have $Rc_1 \subseteq R(\beta c_1 + \delta c_i) + Rc_i$. This implies that β is a unit. Similarly δ is a unit. Replace c_i by $\beta\delta^{-1}c_i$ as a generator of αRb_i . Thus assume

$$\alpha R(b_1 + b_i) = R(c_1 + c_i) \qquad i \geq 2 .$$

(c) If β is in R then a similar argument will show that

$$\alpha R(b_1 + \beta b_2) = R(\delta c_1 + \varepsilon c_2)$$

where δ is a unit. Define

$$\sigma : R \to R$$

by

$$\sigma(\beta) = \delta^{-1}\varepsilon .$$

It is clear that σ is injective since α is injective and that $\sigma(0) = 0$ and $\sigma(1) = 1$.

The object of this proof is to show $\sigma : R \to R$ is a ring automorphism and $\alpha R(\sum a_i b_i) = R(\sum \sigma(a_i)c_i)$ where $\sum a_i b_i$ is unimodular. Then, if $\phi : V \to V$ is σ-semi-linear defined by $\phi b_i = c_i$, we have $P(\phi) = \alpha$.

(d) Suppose $i \geq 3$ and $\bar{\sigma} : R \to R$ is defined as in (c) by

$$\alpha R(b_1 + \beta b_i) = R(c_1 + \bar{\sigma}(\beta)c_i) .$$

We claim $\sigma(\beta) = \bar{\sigma}(\beta)$ for all β in R and σ is a ring morphism. We have

$$b_1 + ab_2 + a'b_i \in R(b_1 + ab_2) \oplus Rb_i .$$

Thus

$$\alpha R(b_1 + ab_2 + a'b_i) \subseteq R(c_1 + \sigma(a)c_2) \oplus Rc_i$$

and

$$\alpha R(b_1 + ab_2 + a'b_i) = R(b(c_1 + \sigma(a)c_2) + b'c_i) .$$

Similarly

$$\alpha R(b_1 + ab_2 + a'b_i) = R(c(c_1 + \bar{\sigma}(a')c_i) + c'c_2) .$$

Thus

$$\alpha R(b_1 + ab_2 + a'b_i) = R(c_1 + \sigma(a)c_2 + \bar{\sigma}(a')c_i)$$

(use $u(b(c_1 + \sigma(a)c_2) + b'c_i) = c(c_1 + \bar{\sigma}(a')c_i) + c'c_2)$.

Since

$$ab_2 + b_i \in R(b_1 + ab_2 + b_i) \oplus Rb_1$$

we have

$$\alpha R(ab_2 + b_i) = R(b(c_1 + \sigma(a)c_2 + c_i) + cc_1) .$$

But

$$\alpha R(ab_2 + b_i) \subset Rc_2 \oplus Rc_i ,$$

thus $b + c = 0$. Hence $\alpha R(ab_2 + b_i) = R(\sigma(a)c_2 + c_i)$.

We now show σ preserves sums.

For a and a' in R ,

$$\alpha R(b_1 + (a + a')b_2 + b_i) = R(c_1 + \sigma(a + a')c_2 + c_i) .$$

But

$$\begin{aligned}\alpha R(b_1 + (a + a')b_2 + b_i) &\subset \alpha R(b_1 + ab_2) + \alpha R(a'b_2 + b_i) \\ &= R(c_1 + \sigma(a)c_2) + R(\sigma(a')c_2 + c_i) .\end{aligned}$$

Hence $\sigma(a) + \sigma(a') = \sigma(a + a')$.

Next we show σ preserves products and $\sigma = \bar{\sigma}$.

For a and a' in R ,

$$\alpha R(b_1 + aa'b_2 + ab_i) = R(c_1 + \sigma(aa')c_2 + \bar{\sigma}(a)c_i) .$$

But

$$\alpha R(b_1 + aa'b_2 + ab_i) \subset \alpha Rb_1 + \alpha R(a'b_2 + b_i)$$

and

$$\alpha R(b_1 + aa'b_2 + ab_i) = R(bc_1 + b'(\sigma(a')c_2 + c_i)) .$$

Thus $\sigma(aa') = \bar{\sigma}(a)\sigma(a')$. Setting $a' = 1$ we have $\sigma(a) = \bar{\sigma}(a)$ for all a in R . Therefore $\sigma(aa') = \sigma(a)\sigma(a')$. Thus σ is an injective ring morphism.

If α is replaced by α^{-1} we obtain $\hat{\sigma} : R \to R$ with $\alpha^{-1}R(c_1 + bc_2) = R(b_1 + \hat{\sigma}(b)b_2)$. Thus $\sigma\hat{\sigma} = \hat{\sigma}\sigma =$ identity and σ is an automorphism of R .

(e) We now show that if a is unimodular and $a = \sum a_i b_i$ then $\alpha Ra = \alpha R(\sum a_i b_i) = R(\sum \sigma(a_i)c_i)$. We begin by showing for $a_2, \ldots, a_n$ in R ,

$$\alpha R(b_1 + a_2 b_2 + \cdots + a_n b_n) = R(c_1 + \sigma(a_2)c_2 + \cdots + \sigma(a_n)c_n) .$$

From above

$$\alpha R(b_1 + a_2 b_2) = R(c_1 + \sigma(a_2)c_2) .$$

The proof is by induction. Assume

$$\alpha R(b_1+a_2b_2+\cdots+a_{n-1}b_{n-1}) = R(c_1+\sigma(a_2)c_2+\cdots+\sigma(a_{n-1})c_{n-1}) .$$

Since

$$\alpha R(b_1+a_2b_2+\cdots+a_nb_n) \subset \alpha R(b_1+a_2b_2+\cdots+a_{n-1}b_{n-1}) \oplus \alpha Rb_n$$

we obtain

$$\alpha R(b_1+a_2b_2+\cdots+a_nb_n) = R[b(c_1+\sigma(a_2)c_2+\cdots+\sigma(a_{n-1})b_{n-1})+b'c_n] .$$

On the other hand,

$$\alpha R(b_1+a_2b_2+\cdots+a_nb_n) \subset \alpha R(b_1+a_nb_n) \oplus \alpha Rb_2 \oplus \cdots \oplus \alpha Rb_{n-1} .$$

Equating coefficients yields $b' = b\sigma(a_n)$ giving the desired result.

If $a_2,\ldots,a_n$ are in R with $a_2b_2 + \cdots + a_nb_n$ unimodular, then

$$\alpha R(a_2b_2 + \cdots + a_nb_n) \subset \alpha R(b_1 + a_2b_2 + \cdots + a_nb_n) \oplus \alpha Rb_1 .$$

Then by the above

$$\alpha R(a_2b_2+\cdots+a_nb_n) = R[b(c_1+\sigma(a_2)c_2+\cdots+\sigma(a_n)c_n)+b'c_1] .$$

Also

$$\alpha R(a_2b_2 + \cdots + a_nb_n) \subset Rc_2 \oplus \cdots \oplus Rc_n .$$

Combining these equations

$$\alpha R(a_2b_2 + \cdots + a_nb_n) = R(\sigma(a_2)c_2 + \cdots + \sigma(a_n)c_n) .$$

Using an argument similar to the above, for any i $(2 \le i \le n)$,

$$\alpha R(b_i + a_1b_1 + \cdots + a_{i-1}b_{i-1} + a_{i+1}b_{i+1} + \cdots + a_nb_n)$$
$$= R(c_i + \sigma(a_1)c_1 + \cdots + \sigma(a_n)b_n) .$$

(f) Finally, let $e = \sum a_ib_i$ be unimodular. We claim

$$\alpha R(a_1b_1 + \cdots + a_nb_n) = R(\sigma(a_1)c_1 + \cdots + \sigma(a_n)c_n) .$$

For $i = 1,2,3,\ldots,n$ let

$$d_i = a_1b_1 + \cdots + a_{i-1}b_{i-1} + a_{i+1}b_{i+1} + \cdots + a_nb_n .$$

Then

$$\alpha Re \subset \alpha Rb_i + \alpha R(b_i + d_i) .$$

Thus $\alpha Re = Rf$ where

$$\begin{aligned} f &= \beta_1\sigma(a_1)c_1 + \alpha_1\sigma(a_2)c_2 + \alpha_1\sigma(a_3)c_3 + \cdots \\ &= \alpha_2\sigma(a_1)c_1 + \beta_2\sigma(a_2)c_2 + \alpha_2\sigma(a_3)c_3 + \cdots \\ &= \alpha_3\sigma(a_1)c_1 + \alpha_3\sigma(a_2)c_2 + \beta_3\sigma(a_3)c_3 + \cdots \quad . \end{aligned}$$

Equating coefficients

$$\beta_1\sigma(a_1)\sigma(a_2) = \alpha_3\sigma(a_1)\sigma(a_2) = \alpha_1\sigma(a_1)\sigma(a_2)$$

and for each $i \ge 3$

$$\beta_1\sigma(a_1)\sigma(a_i) = \alpha_2\sigma(a_1)\sigma(a_i) = \alpha_1\sigma(a_1)\sigma(a_i) .$$

Since $e = \sum a_i b_i$ is unimodular, there exist $k_1, \ldots, k_n$ with $\sum \sigma(a_i) k_i = 1$. Set

$$d = \beta_1 \sigma(a_1) k_1 + \alpha_1 \sigma(a_2) k_2 + \cdots + \alpha_1 \sigma(a_n) k_n .$$

Then from the above equations

$$d\sigma(a_1) = \beta_1 \sigma(a_1)(\sum \sigma(a_i) k_i) = \beta_1 \sigma(a_1)$$

and

$$d\sigma(a_i) = \alpha_1 \sigma(a_i) \quad \text{for} \quad i \geq 2 .$$

Hence $f = d(\sum \sigma(a_i) c_i)$. But f is unimodular. Hence d is a unit.

This completes the proof.

We examine the uniqueness of representation of the projectivity by a semi-linear isomorphism.

Suppose ϕ_1 and ϕ_2 are σ_1 and σ_2-semi-linear, respectively, and $P(\phi_1) = P(\phi_2)$. Then

$$P(\phi_2^{-1} \phi_1) : P(V) \to P(V)$$

fixes lines. Thus if e is unimodular $\phi_2^{-1} \phi_1 e = \lambda e$ for λ a unit in R . Let e , f and g be unimodular and suppose

$$\phi_2^{-1}\phi_1 : \begin{cases} e \to \lambda e \\ f \to \beta f \\ g \to \delta g \end{cases} .$$

Further suppose e , f and g are chosen so that $g = e + f$. Then

$$\phi_2^{-1}\phi_1(e + f) = \phi_2^{-1}\phi_1(e) + \phi_2^{-1}\phi_1 f \text{ , i.e.,}$$

$$\delta(e + f) = \lambda e + \beta f .$$

Hence $\delta = \lambda = \beta$. Thus $\phi_2^{-1}\phi_1(e) = \lambda e$ where λ is fixed for all unimodular vectors. Thus

$$\begin{aligned} \phi_1(e) &= \phi_2(\lambda e) \\ &= \sigma_2(\lambda)\phi_2(e) . \end{aligned}$$

Let $\bar{\lambda} = \sigma_2(\lambda)$. Note $\bar{\lambda}$ is a unit and $\phi_1 = \bar{\lambda}\phi_2$. It is easy to see that $\sigma_1 = \sigma_2$.

(I.14) PROPOSITION. Let V be an R-space of dimension ≥ 3 . Let ϕ_i be a σ_i-semi-linear isomorphism for $i = 1,2$. If $P(\phi_1) = P(\phi_2)$ then $\sigma_1 = \sigma_2$ and $\phi_1 = \lambda\phi_2$ where λ is a unit of R .

(E) SOME GROUP THEORY

Throughout these notes we will be concerned with the action of a classical linear group over a local ring on itself, on a suitable set of vectors or on its R-space. We state the basic group theoretic terminology and ideas involved in this action.

Let E be a set and G be a group. Then G is said to <u>act on</u> E <u>as a transformation group</u> if there exists a map

$$G \times E \to E$$
$$\langle\sigma, s\rangle \to \sigma s$$

satisfying

(a) $1s = s$ for 1 (identity) in G and s in E.

(b) $\sigma(\beta s) = (\sigma\beta)s$ for σ and β in G and s in E.

Two elements s and t of E are G-<u>equivalent</u> if there is a σ in G with $\sigma s = t$.

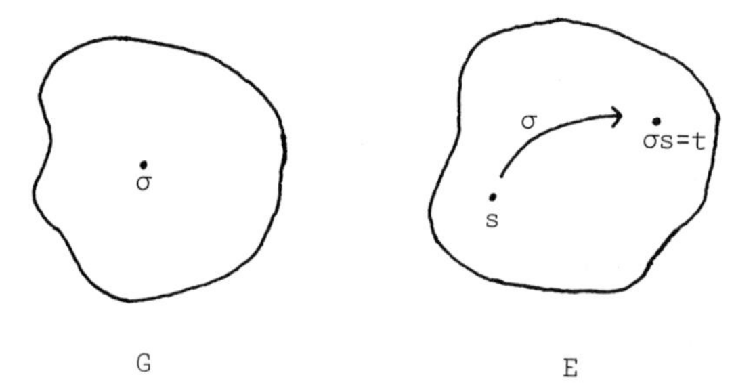

This is an equivalence relation on E. The equivalence classes of E under G-equivalence are the <u>orbits of</u> E <u>under</u> G. If there is only one orbit in E, i.e., E itself, then G is said to be <u>transitive</u> on E.

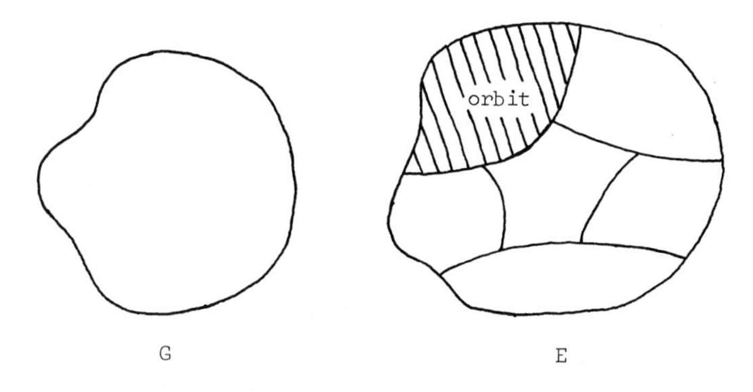

Observe G is transitive as a transformation group on each of its orbits in E. If s is in E, then the stablizer of s is

$$I(s) = \{\sigma \text{ in } G \mid \sigma s = s\} .$$

Note that the stablizer of s is a subgroup of G, denoted $I(s) \leq G$.

When we restrict our attention to a single orbit, G is transitive and the following basic result describes the action of G on the orbit. This theorem may be found in Sah's Abstract Algebra.

(I.15) THEOREM. (Fundamental Theorem of Transformation Groups) Let G act as a transitive transformation group on a set E. Let s be a point in E. Then

(a) $I(\sigma s) = \sigma I(s)\sigma^{-1}$ for all σ in G, i.e., the conjugate of the stablizer $I(s)$ of s by σ is the stablizer of σs.

(b) Let $H = I(s)$. Then G acts transitively on the set of cosets G/H of H in G by

$$G \times G/H \to G/H$$
$$\langle\sigma, \beta H\rangle \to \sigma\beta H \quad \text{(left translation)} .$$

Further, $I(\beta H) = \beta H \beta^{-1}$.

(c) The map

$$G/H \xrightarrow{\tau} E$$

by

$$\tau : \sigma H \to \sigma s$$

is a bijection. Further, the following diagram is commutative.

$$\begin{array}{ccc} G \times G/H & \longrightarrow & G/H \\ {\scriptstyle id \times \tau}\downarrow & & \downarrow{\scriptstyle \tau} \\ G \times E & \longrightarrow & E \end{array} .$$

The proof of the above theorem is straightforward. The above result basically models the action of G on E by replacing E by a set of cosets $G/I(s)$. The action of G on E becomes left translation on $G/I(s)$.

An important example of transformation groups lies in the context of semi-direct products.

Let H and K be groups. Suppose there is a group morphism

$$\Theta : H \to Aut(K)$$

taking H into the group automorphisms $\mathrm{Aut}(K)$ of K. If $h \to \Theta(h)$ denote $\Theta(h)(k)$ by k^h. Suppose further

(a) $k^1 = k$,

(b) $(k^{h_1})^{h_2} = k^{h_1 h_2}$.

Thus H acts as a transformation group on K. Form the product (set theoretic)

$$H \times K = \{\langle h,k\rangle \mid h \text{ is in } H, k \text{ is in } K\} ,$$

and let

$$\langle h_1,k_1\rangle\langle h_2,k_2\rangle = \langle h_1h_2,k_1^{h_2}k_2\rangle .$$

Under this product $H \times K$ is a group called the _semi-direct product of_ H _and_ K.

Observe if $\Theta : H \to \mathrm{Aut}(K)$ is given by $\Theta(h) = 1$, then the semi-direct product is the ordinary direct product of H and K.

The following result is straightforward.

(I.16) THEOREM. (Semi-direct Products) Let G be a group. Let H and K be subgroups of G. Then the following are equivalent:

(a) G is the semi-direct product of H and K for some $\Theta : H \to \mathrm{Aut}(K)$.

(b) (1) $K \triangleleft G$ (K is normal in G)

(2) $HK = G$

(3) $H \cap K = 1$

(c) There is an exact sequence

$$1 \longrightarrow K \xrightarrow{\text{inj}} G \xrightarrow{\sigma} H' \longrightarrow 1$$

which splits with splitting morphism $\delta : H' \to G$ such that $\sigma\delta = id_{H'}$ and $\delta(H') = H$.

We apply the above discussion to the previous sections.

The sequence

$$1 \longrightarrow SL(V) \longrightarrow GL(V) \xrightarrow{\det} R^* \longrightarrow 1$$

splits with splitting morphism $d_n(\lambda) \leftarrow \lambda$. Thus $GL(V)$ is a semi-direct product of $SL(V)$ and R^*. Precisely,

$$GL(V) \simeq \{\langle\sigma,\delta\rangle \mid \sigma \text{ in } SL(V) , \delta \text{ in } R^*\}$$

where

$$\langle\sigma,\delta\rangle\langle\sigma_1,\delta_1\rangle = \langle\sigma\sigma_1,(\sigma_1^{-1}\delta\sigma_1)\delta_1\rangle$$

(here $\delta \equiv d_n(\delta)$).

Let

$$SGL(V) = \{\phi : V \to V \mid \phi \text{ is invertible semi-linear}\} .$$

Then

$$1 \to GL(V) \to SGL(V) \to Aut(R) \to 1$$

under

$$\sigma \to \langle \sigma, 1 \rangle$$
$$\langle \sigma, \beta \rangle \to \beta$$
$$\langle 1, \beta \rangle \leftarrow \beta$$

is a split exact sequence. Thus $SGL(V)$ is a semi-direct product of $GL(V)$ and $Aut(R)$.

Let $SP(V)$ denote the <u>group of projectivities</u> on V and assume $\dim(V) \geq 3$. By the Fundamental Theorem of Projective Geometry, there is a surjective group morphism

$$SGL(V) \xrightarrow{P} SP(V)$$
$$\phi \longrightarrow P(\phi) \ .$$

The kernel of P is R^* . Thus we have a natural exact sequence

$$1 \to R^* \to SGL(V) \to SP(V) \to 1 \ .$$

Let $PGL(V) = GL(V)/Center(GL(V))$ denote the <u>projective general linear group</u>. Recall

$$Center(GL(V)) = \{\beta \mid \beta\sigma = \sigma\beta \text{ for all } \sigma \text{ in } GL(V)\}$$
$$= \{uI \mid u \text{ in } R^*\} \ .$$

The second statement is shown by realizing that β commutes with all transvections and this forces β to be a scalar multiple of the identity. Hence $\text{Center}(GL(V)) \simeq R^*$ and there is a natural exact sequence

$$1 \to R^* \to GL(V) \to PGL(V) \to 1 \ .$$

Since each σ in $GL(V)$ also induces a projectivity and σ and β induce the same projectivity if and only if $\sigma \equiv \beta \mod R^*$, we have $PGL(V) \leq SP(V)$.

This discussion is summarized in the following diagram:

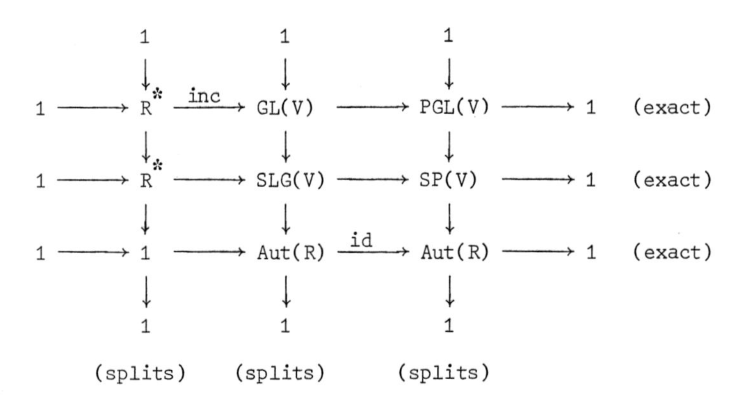

Theorem (I.12) characterized the automorphisms of End(V) as conjugations by semi-linear isomorphisms. Suppose

$$\wedge_{\phi_1}(\sigma) = \phi_1^{-1}\sigma\phi_1$$

and

$$\Lambda_{\phi_2}(\sigma) = \phi_2^{-1}\sigma\phi_2$$

where $\phi_1 \longleftrightarrow \langle P_1,\sigma_1\rangle$ and $\phi_2 \longleftrightarrow \langle P_2,\sigma_2\rangle$.

If $\Lambda_{\phi_1} = \Lambda_{\phi_2}$ then $\phi_1^{-1}\sigma\phi_1 = \phi_2^{-1}\sigma\phi_2$ or

$$(\phi_2\phi_1^{-1})\sigma = \sigma(\phi_2\phi_1^{-1})$$

for all σ in End(V) .

It is easy to check that $\Lambda_{\phi_1} = \Lambda_{\phi_2}$ if and only if $P_1 = \lambda P_2$ where λ is in R^* and $\sigma_1 = \sigma_2$. Further the map

$$\mathrm{Aut}(\mathrm{End}(V)) \to \mathrm{SP}(V)$$

by

$$\Lambda_\phi \to \phi$$

is a group isomorphism.

(I.17) PROPOSITION. The automorphism group of End(V) is isomorphic to a semi-direct product of PGL(V) and Aut(R) .

(F) THE RING OF ENDOMORPHISMS OF V

Section (C) contained a few remarks on relations between V , $V^* = \mathrm{Hom}_R(V,R)$, R and $\mathrm{End}_R(V)$. This section provides a more detailed discussion of these concepts. This material is not essential to the remainder of the monograph.

Setting:

(a) R a local ring with maximal ideal m and residue field $k = R/m$.

(b) V an R-space of dimension n .

(c) $V^* = \mathrm{Hom}_R(V,R)$ the dual space of V .

(d) $E = \mathrm{End}_R(V)$ the endomorphism ring of V .

Observe V is a left E-module under $\sigma \cdot v = \sigma(v)$ for σ in E and v in V . Hence, V is a left (E,R)-module and $r(\sigma \cdot v) = \sigma \cdot (rv)$ for σ in E , r in R and v in V .[1]

We have already observed that V^* is an R-space. Since R is commutative, V^* may be viewed as a right R-space or a left R-space (depending on preference). We choose to view V^* as a right R-space in this section. In addition, V^* is naturally a right E-module under $f \cdot \sigma$ given by $(f \cdot \sigma)(v) = f(\sigma(v))$ for f in V^* , σ in E and v in V . Consequently, V^* is a right (E,R)-module.

---

[1] By an (E,R)-module V we mean a V is an E-module and an R-module and the E and R actions are compatible on V .

Select a unimodular b in V. Then $V = Rb \oplus W$. Let $\Pi_b : V \to Rb$ and $\Pi_W : V \to W$ be the natural projections and $\lambda_b : Rb \to V$ and $\lambda_W : W \to V$ be the natural injections. Relative to the decomposition $V = Rb \oplus W$, an R-morphism $\sigma : V \to V$ may be identified with a matrix

$$\mathrm{Mat}(\sigma) = \begin{bmatrix} \alpha & \beta \\ \delta & \eta \end{bmatrix}$$

where $\alpha = \Pi_b\sigma\lambda_b$ is in $\mathrm{Hom}_R(Rb,Rb)$, $\beta = \Pi_b\sigma\lambda_W$ is in $\mathrm{Hom}_R(W,Rb)$, $\delta = \Pi_b\sigma\lambda_b$ is in $\mathrm{Hom}_R(Rb,W)$ and $\eta = \Pi_W\sigma\lambda_W$ is in $\mathrm{Hom}_R(W,W)$. Conversely, given α, β, δ and η of the above type, we may construct a unique element of E. Indeed, $\sigma \to \mathrm{Mat}(\sigma)$ is a ring isomorphism.

If v is in V then let $v = rb + w$ and select α in $\mathrm{Hom}_R(Rb,Rb)$ by $\alpha : b \to rb$ and δ in $\mathrm{Hom}_R(Rb,W)$ by $\delta : b \to w$. Then

$$\begin{bmatrix} \alpha & 0 \\ \delta & 0 \end{bmatrix}$$

carries b to v. Hence, the cyclic E-module Eb is all of V, i.e., $V = Eb$.

There is a natural E-morphism $E\Pi_b \to Eb = V$ given by $\sigma\Pi_b \to b$. It is easy to check that this mapping is well-defined and bijective. Thus, as E-modules, $V \simeq E\Pi_b$. Since $1 = \Pi_b + \Pi_w$ and Π_b and Π_w are orthogonal idempotents, that is, $\Pi_b^2 = \Pi_b$, $\Pi_w^2 = \Pi_w$ and $0 = \Pi_b\Pi_w = \Pi_w\Pi_b$ in E , then $E = E\Pi_b \oplus E\Pi_w$ as a left E-module. Hence ${}_EV$ is isomorphic to a direct summand of ${}_EE$. By Section (I.B) we call ${}_EV$ an E-subspace of the E-space ${}_EE$. However, contrary to the earlier setting, e.g., (I.5), ${}_EV$ is itself not an E-space, i.e., ${}_EV$ is not E-free. Thus, ${}_EV$ is a proper "projective" E-module.

(I.18) PROPOSITION.

(a) $R \simeq \mathrm{End}_E(V)$ (as rings) under $r \to \rho_r$ where $\rho_r(v) = rv$ for r in R .

(b) $V^* \simeq \mathrm{Hom}_E(V,E)$ (as right E-modules) under $f \to \beta_f$ where $\beta_f(v) = f(\ )v$ for f in V^* .

(c) $E^0 \simeq \mathrm{Hom}_R(V^*,V^*)$ (as rings) under $\sigma \to \tau_\sigma$ where $\tau_\sigma(f) = f\sigma$ for σ in E^0 . (E^0 denotes the "opposite" ring of E , i.e., as an Abelian group $E = E^0$, however multiplication $*$ in E^0 is given by $x * y = yx$.)

The proofs of the above are straightforward.

Suppose A is an ideal of R . Then

$$AV = \{\textstyle\sum_f a_i v_i \mid a_i \in A, v_i \in V\}$$

($\sum_f$ denotes finite sum) is an E-submodule of ${}_EV$. On the other hand, if T is an E-submodule of ${}_E V$, then

$$\mathcal{T}_R(T) = \{\sum_f f_i(t_i) \mid f_i \in V^*, t_i \in T\}$$

is an ideal of R . This ideal $\mathcal{T}_R(T)$ is called the R-trace of T .

For the ideal A of R , consider the maps

$$A \to AV \to \mathcal{T}_R(AV) .$$

If $t = \sum_f a_i v_i$ is in AV with a_i in A and v_i in V then $\sigma(t) = \sum a_i \sigma(v_i)$ is in A for σ in V^* . Hence, $\mathcal{T}_R(AV) \subseteq A$. On the other hand, if $\rho_b : V \to R$ by $\rho_b(b) = 1$ and $\rho_b(W) = 0$, then $A = \rho_b(Ab) \subseteq \mathcal{T}_R(AV)$. Thus $A = \mathcal{T}_R(AV)$.

If T is an E-submodule of ${}_E V$, we have the following maps

$$T \to \mathcal{T}_R(T) \to \mathcal{T}_R(T)V$$

from E-submodules to ideals of R to E-submodules. We claim $T = \mathcal{T}_R(T)V$. If $\{b_1,\ldots,b_n\}$ is a basis for V with dual basis $\{b_1^*,\ldots,b_n^*\}$ for V^* then, for $t = \sum r_i b_i$ in T , we have $b_i^* t = r_i$ in $\mathcal{T}_R(T)$. Hence $r_i b_i$ is in $\mathcal{T}_R(T)V$ and therefore $T \subset \mathcal{T}_R(T)V$.

It remains to show $\mathcal{T}_R(T)V \subset T$.

Observe if f is in V^* and v is in V then $[f,v] : V \to V$ by $[f,v](w) = f(w)v$ is an R-linear map. That is, $[\ ,\ ] : V^* \times V \to E$.

Suppose $t = \sum a_i b_i$ is in $\mathcal{T}_R(T)V$. We may assume a_i is in $\mathcal{T}_R(T)$ for $1 \le i \le n$. Then $a_i = \sum_j f_{ij}(t_{ij})$ for t_{ij} in T and f_{ij} in V^*. Then

$$t = \sum_i a_i b_i = \sum_i \sum_j f_{ij}(t_{ij}) b_i$$
$$= \sum_i \sum_j [f_{ij}, b_i] t_{ij} .$$

But the $[f_{ij}, b_i]$ are in E and the t_{ij} come from T — an E-submodule of V. Hence t is in T. That is, $\mathcal{T}_R(T)V \subset T$.

(I.19) THEOREM. The above mappings

$$A \to AV$$

and

$$\mathcal{T}_R(T) \leftarrow T$$

induce a lattice isomorphism between

$$\{\text{Ideals of } R\} \leftrightarrow \{\text{E-submodules of } {}_EV\} .$$

Proof. The previous discussion shows that the maps induce a bijection between the sets. It is obvious that they preserve partial order and easy to show sums and intersections are preserved.

We pause to single out two natural pairings which appear in the discussion prior to (I.19).

First, the map (,)

$$(\ , \ ) : V^* \times V \to R$$

by

$$\langle f,v\rangle \to (f,v) = f(v) \ .$$

Second, the map [,]

$$[\ , \ ] : V^* \times V \to E$$

by

$$\langle f,v\rangle \to [f,v]$$

where $[f,v](w) = f(w)v$ for w in V . Note $[f,v](w) = (f,w)v$.
These pairings are balanced and induce

$$(\ , \ ) : V^* \otimes_E V \to R$$

and

$$[\ , \ ] : V^* \otimes_R V \to E$$

which are respectively, (R,R) - and (E,E) – module isomorphisms. However, we will not make use of the tensor product in this monograph.

Suppose A is a right ideal of E . Consider AV . Since ${}_EV \simeq E\Pi_b$ (as E-modules) where $E = E\Pi_b \oplus E\Pi_w$, it is clear $AV \simeq A\Pi_b$ (as an R-module) and $A = A\Pi_b \oplus A\Pi_w$. The map

$$A \to AV$$

associates right ideals of E to R-submodules of V. On the other hand, if Q is an R-submodule of V, we define the E-trace of Q

$$\mathcal{T}_E(Q) = \{\sum_f [g_i,q_i] \mid g_i \in V^*, q_i \in Q\} .$$

Observe, for σ in E, g in V^* and q in Q, $[g,q]\sigma = [g\sigma,q]$ is in $\mathcal{T}_E(Q)$. Hence $\mathcal{T}_E(Q)$ is a right ideal of E.

The map

$$Q \to \mathcal{T}_E(Q)$$

associates R-submodules of V to right ideals of E.

It is easy to see

$$A \to AV \to \mathcal{T}_E(AV) = A .$$

On the other hand, for Q an R-submodule of V consider $\mathcal{T}_E(Q)V$. If x is in $\mathcal{T}_E(Q)V$ then

$$\begin{aligned} x &= \sum [g_i,q_i]v_i \\ &= \sum g_i(v_i)q_i \end{aligned}$$

is in Q for g_i in V^*, q_i in Q and v_i in V. Thus $\mathcal{T}_E(Q)V \subseteq Q$. Conversely, if q is in Q, then $q = [f,q]b$ is in $\mathcal{T}_E(Q)V$. Hence

$$Q \to \mathcal{T}_E(Q) \to \mathcal{T}_E(Q)V = Q .$$

It is straightforward to show these maps preserve the lattice structure.

(I.20) THEOREM. The above mappings

$$A \to AV$$

and

$$T_E(Q) \leftarrow Q$$

induce a lattice isomorphism between

$$\{\text{right ideals of } E\} \leftrightarrow \{R\text{-submodules of } {}_RV\} .$$

We could also seek to determine in some fashion the left ideals of E . For this recall V^* is a right E-module and there is a natural R-isomorphism

$$V \to V^{**} = (V^*)^*$$

given by

$$x \to (\ ,x) .$$

Thus, in the previous discussion, replace V by V^* and V^* by V^{**} , respectively, and parallel the arguments.

(I.21) THEOREM. There is a lattice isomorphism between the lattice of left ideals of E and the lattice of R-submodules of V^* .

We now examine a subclass of the left and right ideals of E — those ideals which occur as annihilators.

Let S be a subset of V, then the <u>orthogonal complement</u> $S^{\perp}$ <u>of</u> S <u>in</u> V^* is the set of all g in V^* satisfying $(g,S) = 0$, i.e., $(g,s) = 0$ for all s in S. If $\langle S \rangle$ denotes the R-submodule of V generated by S then $\langle S \rangle^{\perp} = S^{\perp}$. If T is a subset of V^*, then $T^{\perp}$ is the set of all v in V with $(T,v) = 0$. An R-submodule S of V (or V^*) is <u>closed</u> if $S^{\perp\perp} = (S^{\perp})^{\perp}$ is equal to S. Clearly $S \subseteq S^{\perp\perp}$ and $S^{\perp\perp\perp} = S^{\perp}$.

Let B be any subset of E. Define

$$R(B) = \{\sigma \text{ in } E \mid B\sigma = 0\}$$

to be the <u>right annihilator of</u> B and

$$L(B) = \{\sigma \text{ in } E \mid \sigma B = 0\}$$

to be the <u>left annihilator of</u> B. If $\bar{B}$ is the left (right) ideal generated by B then $R(B) = R(\bar{B})$ ($L(B) = L(\bar{B})$). Further, $R(B)$ is a right ideal and $L(B)$ is a left ideal in E. If A is a right (left) ideal of E and $A = R(B)$ ($A = L(B)$) for some $B \subset E$ then A is called a <u>right</u> (<u>left</u>) <u>annulet</u>.

The set of right annulets in E is a lattice where

$$A \vee B = R(L(A + B))$$

and

$$A \wedge B = A \cap B .$$

Further, the map $A \to L(A)$ is a lattice anti-isomorphism between the right annulets of E and the lattice of left annulets of E.

Let S be a subset of V. Define

$$\underline{L}(S) = \{\sigma \text{ in } E \mid \sigma S = 0\}$$

and

$$\underline{R}(S) = \{\sigma \text{ in } E \mid \sigma V \subseteq S\} .$$

Finally, for $B \subseteq E$, let

$$N(B) = \{v \text{ in } V \mid Bv = 0\}$$

and

$$BV = \{\sigma v \mid \sigma \text{ in } B \text{ and } v \text{ in } V\} .$$

Call $N(B)$ the <u>kernel</u> of B and BV the <u>image</u> of B.

<u>(I.22)</u> <u>LEMMA</u>. $\underline{R}(S)V = S$ where S is an R-submodule of V.

<u>Proof</u>. Clearly $\underline{R}(S)V \subseteq S$. Let s be in S and $V = Rb \oplus W$. Define σ in E by $\sigma b = s$ and $\sigma|_W = 0$. Then $\sigma V = Rs \subset S$ and σ is in $\underline{R}(S)$. Also $s = \sigma b$. Thus, s is in $\underline{R}(S)V$ and $S \subseteq \underline{R}(S)V$.

The next lemma is straightforward.

<u>(I.23)</u> <u>LEMMA</u>.

(a) For each subset $S \subseteq V$, we have $S \subseteq N(\underline{L}(S))$.

(b) For each subset $S \subseteq V$, we have $\underline{L}(S)\underline{R}(S) = 0$.

(c) For each subset $B \subseteq E$, $\mathcal{R}(B) = \underline{R}(N(B))$.

(d) For each subset $B \subseteq E$, $\mathcal{L}(B) = \underline{L}(BV)$.

<u>(I.24)</u> <u>THEOREM</u>.

(a) Let S be an R-submodule of V . Then $\mathcal{L}(\underline{R}(S)) = \underline{L}(S)$.

(b) Let $J = \mathcal{L}(B)$ for $B \subseteq E$. Then $J = \underline{L}(N(J))$.

(c) Let $J = \mathcal{R}(B)$ for $B \subseteq E$. Then JV is an R-submodule of V and $J = \underline{R}(JV)$.

<u>Proof</u>. To show (a), use (I.23) then (I.22) and

$$\mathcal{L}(\underline{R}(S)) = \underline{L}(\underline{R}(S)V) = \underline{L}(S) \quad .$$

To show (b), let $J = \mathcal{L}(B)$. Then

$$\mathcal{R}(J) = \mathcal{R}(\mathcal{L}(B))$$

and

$$\mathcal{L}(\mathcal{R}(J)) = \mathcal{L}(\mathcal{R}(\mathcal{L}(B))) = \mathcal{L}(B) \ .$$

Thus

$$\begin{aligned} J &= \mathcal{L}(\mathcal{R}(J)) \\ &= \underline{L}(\mathcal{R}(J)V) && \text{by (I.23)(d)} \\ &= \underline{L}(\underline{R}(N(J))V) && \text{by (I.23)(c)} \\ &= \underline{L}(N(J)) && \text{by (I.22)} \quad . \end{aligned}$$

To show (c), let $J = \mathcal{R}(B)$. Then

$$J = \underline{R}(N(B))$$

by (I.23)(c). Thus

$$\begin{aligned} JV &= \underline{R}(N(B))V \\ &= N(B) \end{aligned}$$

by (I.22). Thus

$$\begin{aligned} \underline{R}(JV) &= \underline{R}(N(B)) \\ &= \mathcal{R}(B) \qquad \text{by (I.23)(b)} \\ &= J \quad . \end{aligned}$$

The purpose of these results is to relate the right and left annulets in E to closed submodules of V . Preliminary to this we characterize the closed submodules.

<u>(I.25)</u> <u>THEOREM</u>. Let S be an R-submodule of V . Then S is closed if and only if $S = N(\underline{L}(S))$.

<u>Proof</u>. Suppose $S = N(\underline{L}(S))$ and $B = \underline{L}(S)$. Then $BS = 0$.

If σ is in E , we may define the <u>adjoint</u> or <u>transpose</u> σ^* of σ in $\text{End}_R(V^*)$ by requiring σ^* satisfy

$$(g,\sigma v) = (g\sigma^*,v)$$

for all g in V^* and v in V.

Thus, since $BS = 0$, $V^*B^* \subseteq S^\perp$. Suppose x is in $S^{\perp\perp}$. Then $0 = (S^\perp,x)$. Hence $0 = (V^*B^*,x) = (V^*,Bx)$. But, for any v in V, $(V^*,v) = 0$ implies $v = 0$. Therefore, $Bx = 0$. Thus x is in $N(B) = N(\underline{L}(S)) = S$. That is, $S^{\perp\perp} \subseteq S$. Clearly $S \subseteq S^{\perp\perp}$. Hence $S = S^{\perp\perp}$ and S is closed.

Conversely, suppose $S = S^{\perp\perp}$. By (I.23) $S \subset N(\underline{L}(S))$. Thus, it is necessary to show $N(\underline{L}(S)) \subseteq S$.

If $S = V$ this is trivial. If $S \neq V$ then there is a unimodular element e which is in V but not in S. Select any r not in S. Then, there is an h in $S^\perp$ with $(h,r) \neq 0$ (for otherwise, $(S^\perp,r) = 0$ and r would be in $S^{\perp\perp} = S$). Define $\rho : V \to V$ by $\rho(v) = (h,v)e$ for v in V. Then $\rho S = 0$. Further, $\rho(r) \neq 0$ since e is unimodular. Thus $\underline{L}(S)r \neq 0$ and r is not in $N(\underline{L}(S))$. By contrapositive, $N(\underline{L}(S)) \subseteq S$.

<u>(I.26)</u> <u>COROLLARY</u>. Let S be a subspace of V. Then S is closed.

<u>Proof</u>. We show $N(\underline{L}(S)) \subseteq S$. Suppose $V = S \oplus T$ and a is not in S. Then $a = s + t$ where s is in S and t is in T with $t \neq 0$. Let $\{f_\lambda\},\{x_\lambda\}$ be a dual basis for T, i.e., $f_\lambda(x_\beta) = \delta_{\lambda\beta}$, and extend f_λ to V by

$f_\lambda|_S = 0$. Then $t = \sum (f_\lambda, t)x_\lambda$. Set $\sigma = \sum [f_\lambda, x_\lambda]$. Then $\sigma S = 0$ and $\sigma(a) = t \neq 0$. Hence σ is in $\underline{L}(S)$ but $\underline{L}(S)a \neq 0$. Hence, a is not in $N(\underline{L}(S))$. Consequently, $N(\underline{L}(S)) \subseteq S$.

<u>(I.27)</u> <u>COROLLARY</u>. If S is closed in V , then

$$\underline{R}(S) = \mathcal{R}(\underline{L}(S)) .$$

<u>Proof</u>. $\underline{R}(S) = \underline{R}(N(\underline{L}(S)))$ by (I.25)

$= \mathcal{R}(\underline{L}(S))$ by (I.23)(c) .

Let

$$\begin{aligned} \mathcal{C} &= \{S \subseteq V \mid S \text{ is a closed R-submodule}\} \\ \mathcal{A}^\perp &= \{J \subseteq E \mid J \text{ is a right annulet}\} \\ {}^\perp\mathcal{A} &= \{H \subseteq E \mid H \text{ is a left annulet}\} \end{aligned} .$$

<u>(I.28)</u> <u>THEOREM</u>.

(a) The mappings

$$J \rightarrow \mathcal{L}(J)$$
$$\mathcal{R}(H) \leftarrow H$$

determine a lattice anti-isomorphism between $\mathcal{A}^\perp$ and ${}^\perp\mathcal{A}$.

(b) The mappings

$$J \rightarrow JV$$
$$\underline{R}(S) \leftarrow S$$

determine a lattice isomorphism between $A^{\perp}$ and $\mathcal{C}$.

(c) The mappings

$$H \rightarrow N(H)$$
$$\underline{L}(S) \leftarrow S$$

determine a lattice anti-isomorphism between ${}^{\perp}A$ and $\mathcal{C}$.

(d) Relative to elements of $\mathcal{C}$, the mappings of (a) are given by

$$\underline{R}(S) \leftrightarrow \underline{L}(S)$$

for S in $\mathcal{C}$.

The above result is called the "Three-Cornered Galois Theory of Baer." Pictorially

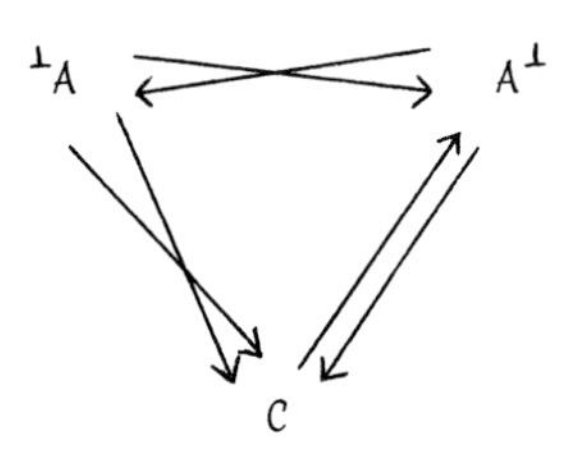

Proof. Part (a) is immediate.

To show (b), suppose S is closed. Then $\underline{R}(S) = R(\underline{L}(S))$ is a right annulet by (I.27). Further, $\mathcal{R}(\underline{L}(S))V = \underline{R}(S)V = S$ by (I.22). Conversely, suppose $J = \mathcal{R}(B)$ is a right annulet. Then

$$J \to JV \to \underline{R}(JV) = J$$

by (I.24)(c). Further, JV is closed. To show this let $J = \mathcal{R}(B) = \underline{R}(N(B))$ by (I.23). Then $JV = \underline{R}(N(B))V = N(B)$ by (I.22). Thus $N(\underline{L}(JV)) = N(\underline{L}(N(B))) = N(B)$ by (I.24).

To show (c), suppose S is closed. Then $\underline{L}(S) = \mathcal{L}(\underline{R}(S))$ is a left annulet by (I.24). Further

$$S \to \underline{L}(S) \to N(\underline{L}(S)) = S$$

by (I.25). Suppose $H = \mathcal{L}(B)$ is a left annulet. Set $S = N(H)$. Then

$$\begin{aligned} N(\underline{L}(S)) &= N(\underline{L}(N(H))) \\ &= N(H) \qquad \text{by (I.24)} \\ &= S \end{aligned}$$

and S is closed by (I.25). Finally

$$H \to N(H) \to \underline{L}(N(H)) = H$$

by (I.24).

Thus, we have verified the maps are bijections. It is easy to show that they preserve or invert the lattices.

If S and T are R-submodules of V, define

$$\underline{R}(S,T) = \{\sigma \text{ in } E \mid \sigma T \subseteq S\} .$$

Clearly $\underline{R}(S,V) = \underline{R}(S)$ and $\underline{R}(S) \subseteq \underline{R}(S,T)$. There is a natural group morphism

$$\tau : \underline{R}(S,T) \to \mathrm{Hom}_R(V/T,V/S)$$

by

$$\sigma \to \sigma^\tau$$

where $\sigma^\tau : v + T \to \sigma(v) + S$. The kernel of this map is $\underline{R}(S)$. To show τ is surjective, consider

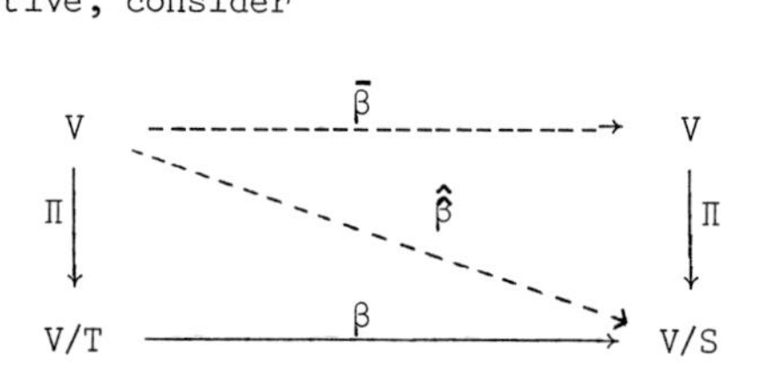

Given $\beta : V/T \to V/S$ we obtain $\hat{\beta} = \beta\Pi : V \to V/S$. Since V is an R-space, $\hat{\beta}$ lifts to $\bar{\beta} : V \to V$. It is easy to see $\bar{\beta}T \subseteq S$ and $\bar{\beta}^\tau = \beta$.

<u>(I.29)</u> <u>PROPOSITION.</u>

(a) There is an exact sequence of R-modules

$$0 \longrightarrow \underline{R}(S) \longrightarrow \underline{R}(S,T) \xrightarrow{\tau} \mathrm{Hom}_R(V/T,V/S) \longrightarrow 0 .$$

(b) If $T = S$, part (a) gives a ring isomorphism $\underline{R}(S,S)/\underline{R}(S) \simeq \mathrm{End}_R(V/S)$. Further, if $T = AV$ for A an ideal of R then this becomes $(R)_n/(A)_n \simeq (R/A)_n$.

(G) DISCUSSION

This chapter provides definitions, terminology and notation for the remainder of the monograph. Sections (A), (B) and (C) are basic and many of the results may be stated for free modules over any commutative ring, e.g., matrix representation, determinant, elementary matrices, etc. A general form of Nakayama's Lemma is given in ([4], p.85) although the sharpest corollaries occur for local rings since here the ring modulo its radical is a field. Generally, if $Rad(R)$ denotes the Jacobson radical of a ring R such results as, "A projective module M is R-free if and only if $M/Rad(R)M$ is $R/Rad(R)$-free" are typical. Observe Corollary (I.4) follows from this. Recall projective modules are direct summands of free modules. The map $M \to M/Rad(R)M$ is a full additive functor from the category of finitely generated projective R-modules to the category of finitely generated projective $R/Rad(R)$-modules which lifts $R/Rad(R)$-isomorphisms to R-isomorphisms. (See [4], p.90.)

Corollary (I.4) illustrates that a direct summand of a free module over a local ring is again free. This is immensely useful in the theory of spaces and their geometries over local rings and permits one to often mimic the arguments for vector spaces over fields for the local ring context. Given a commutative ring R, the determination of precisely when a direct summand of a free R-module is again free has been a widely studied problem. In his paper "Faisceaux algebriques coherents"

Annals of Math. 61(1955), J.-P. Serre conjectured the following: A finitely generated projective module over the ring of polynomials $k[X_1,\dots,X_n]$ in n commuting indeterminates over a field k is free. Recently D. Quillen[1] in "Projective modules over polynomial rings" (soon to appear in *Inventiones Mathematicae*) showed that the Serre conjecture is true if k is either a field or a principal ideal domain! This was proven by using a technique of Murthy which showed any vector bundle over $\mathrm{Proj}(A[X])$ when restricted to the affine line $\mathrm{Spec}(A[T])$ is the base extension of a vector bundle over $\mathrm{Spec}(A)$ where $A = k[X_1,\dots,X_n]$. A key result necessary for the proof was given by Horrocks in 1964. The proof of this conjecture represented the culmination of over 20 years of work by many mathematicians. For introductions to the question of when projectives are free see [25] or [50]. Swan in [17] provides a survey of recent results prior to the Quillen proof.

The result (I.8) is valid for any commutative ring R with trivial idempotents such that the Picard group $\mathrm{Pic}(R)$ has no elements, e.g., rank one projective modules, of order dividing $n = \dim(V)$. An example of this would be any ring having $\mathrm{Pic}(R)$ torsion-free or more specifically any ring having all projective modules free. The key step is that the rank one projective $F_{ij}V$ must be free. If one does not mind the conjugating matrices having elements from a larger ring, then any ring having no non-trivial idempotents may be imbedded in a faithfully flat extension, e.g., a universal splitting ring, which has all rank one projectives free and the result follows in an analogous fashion.

---

[1] Suslin obtained approximately the same results as Quillen at approximately the same time.

Concerning R-algebra automorphisms of $\mathrm{End}_R(V)$, where R is a commutative ring and V a faithfully projective R-module, there is a natural exact sequence

$$1 \longrightarrow \mathrm{Inn}(\mathrm{End}_R(V)) \longrightarrow \mathrm{Aut}(\mathrm{End}_R(V)) \xrightarrow{\alpha} \mathrm{Pic}(R)$$

where $\mathrm{Inn}(\mathrm{End}_R(V))$ denotes the subgroup of inner-automorphisms and $\mathrm{Im}(\alpha)$ is a subgroup of the torsion group of $\mathrm{Pic}(R)$. See [128]. The sequence is generally not exact at $\mathrm{Pic}(R)$. Let $Q(V)$ denote the R-module isomorphism classes of faithfully projective R-modules W satisfying $\mathrm{End}_R(V) \simeq \mathrm{End}_R(W)$. It can be shown that such a W satisfies $\mathrm{End}_R(V) \simeq \mathrm{End}_R(W)$ if and only if $W \simeq E \otimes V$ where E is a rank 1 projective. If [E] is the element of $\mathrm{Pic}(R)$ determined by E , then $[E] \to [E \otimes V]$ gives a group morphism $\mathrm{Pic}(R) \to Q(V)$ and

$$1 \to \mathrm{Inn}(\mathrm{End}_R(V)) \to \mathrm{Aut}(\mathrm{End}_R(V)) \to \mathrm{Pic}(R) \to Q(V) \to 1$$

is exact.

The result (I.10) depends on the fact that over a local ring the group of elementary matrices $E_B(V)$ relative to a given basis B generate the special linear group. This fact $E_B(V) = SL(V)$ is also valid if the ring R is semi-local, Euclidean or a Dedekind domain whose quotient field is a finite extension of the rationals. (See [101].) Recently Cooke [60] has shown this to be true for rings possessing a generalized Euclidean algorithm. However, Bass [5] has exhibited a principal ideal domain R for which $E_B(V) \neq SL(V)$! Indeed, a basic obstruction in the general theory of $GL(V)$ is the lack of understanding of how $E_B(V)$ sits in $GL(V)$. How does one distinguish between $E_B(V)$ and $E_{\bar{B}}(V)$ for bases B and $\bar{B}$ in a group theoretic fashion?[1]

A mapping $\alpha : P(V) \to P(V)$ may be also defined to be a projectivity if

(a) α is bijective.

(b) For lines L, L_1 and L_2, $L \subset L_1 + L_2$ if and only if $\alpha L \subset \alpha L_1 + \alpha L_2$.

This definition was employed in [115] to prove a Fundamental Theorem of Projective Geometry when R was an arbitrary commutative ring (this proof could probably be adapted to R non-commutative also).

We selected the definition given in Section (D) since in application it would require that only lines which form planes need be examined to determine if a given mapping of $P(V)$ is a projectivity. Both definitions imply the Fundamental Theorem, and consequently are equivalent; however, short of reproving this theorem their equivalence is not obvious. A corresponding theorem in dimension 2 [100] is available but only for local rings.

If Section (F) were done properly, one would show that there is a natural categorical equivalence between the category of R-modules and the category of $End_R(V)$-modules. See [4] or [38]. We hesitated to introduce these ideas due to a desire for brevity. The relations between the right and

---

[1] Suslin recently announced that $E_B(V)$ is normal in $GL(V)$ for $|B| \geq 3$ and R any ring!

left annulets of End(V) are well-known; however, the correspondence between these and the closed submodules appears to have escaped general notice. A general treatment for non-finitely generated projective modules relative to density and categorical equivalence is given in [69] and to automorphisms of endomorphism rings in [102]. Further references on non-finitely generated projective modules are [36], [43], [99], [153] and [154]. The literature is modest concerning non-finitely generated projectives. The result (I.29) is related to the concept of an idealizer of a left (or right) ideal and its eigen-ring — see [12].

II. THE GENERAL LINEAR GROUP:
NORMAL SUBGROUPS AND AUTOMORPHISMS

(A) THE GENERAL LINEAR GROUP - INTRODUCTION

Our basic purpose is to determine the normal subgroups of $GL(V)$ and the automorphisms of $GL(V)$ when V is an R-space for a local ring R. We then ask the analogous questions for the orthogonal and symplectic groups.

In this chapter we determine the normal subgroups, automorphisms and simultaneously much of the structure of $GL(V)$. The approach to the normal subgroups is by way of a paper of Klingenberg [89] — basically a coordinate-free technique, i.e., we move lines and hyperplanes and utilize transvections rather than compute with matrices and elementary transvections. A much more general approach which is matrix theoretic is given in [37] by Bass. We use Keenan's [88] work to find the automorphisms of $GL(V)$.

Let R be a local ring with maximal ideal m and residue field $k = R/m$. Let V denote an R-space of dimension n.

If A is an ideal of R then the natural ring morphism $\Pi_A : R \to R/A$ induces an R-morphism $\Pi_A : V \to V/AV$. If $A = R$ we take V/AV to be the zero space. The R-morphism Π_A determines a group morphism

$$\lambda_A : GL(V) \to GL(V/AV) .$$

This is given by

$$(\lambda_A\sigma)\Pi_A = \Pi_A\sigma$$

for all σ in $GL(V)$, i.e., if x is in V then

$$(\lambda_A\sigma)\Pi_A(x) = \Pi_A\sigma(x) .$$

Since R/A is also local, $GL(V/AV)$ is generated by elementary transvections and dilations. Since $R \to R/A$ is surjective, each transvection and dilation in $GL(V/AV)$ has a pre-image in $GL(V)$. Thus λ is a surjective group morphism.

Let x be in V . The _order of_ x

$$O(x) = \text{smallest ideal } A \text{ of } R \text{ such that}$$

$$\Pi_A(x) = 0 .$$

Let σ be in $GL(V)$. The _order of_ σ

$$O(\sigma) = \text{smallest ideal } A \text{ such that}$$

$$\lambda_A(\sigma) \text{ is in the center of } GL(V/AV) .$$

Let G be a subgroup of $GL(V)$, denoted $G \le GL(V)$. The _order of_ G

$$O(G) = \text{smallest ideal } A \text{ such that}$$

$$\lambda_A G \le \text{center of } GL(V/AV) .$$

Let V have a basis $b_1,\ldots,b_n$. Let $x = r_1b_1 + \cdots + r_nb_n$ be in V and let σ in $GL(V)$ satisfy $\sigma b_i = \sum_j a_{ji}b_j$ $(1 \le i \le n)$. Thus $Mat(\sigma) = [a_{ji}]$. Then

(a) $O(x)$ = the ideal $(r_1,\ldots,r_n)$ generated by $r_1,\ldots,r_n$. In particular, x is unimodular if $O(x) = R$.

(b) $O(\sigma)$ = the ideal generated by $\{a_{ji}$ $(j \neq i)$ and $a_{ii} - a_{jj}\}$. Thus

$$[a_{ji}] \xrightarrow{\lambda_{O(\sigma)}} \begin{bmatrix} \bar{r} & & \cdots & & 0 \\ \cdot & \bar{r} & & & \cdot \\ \cdot & & \cdot & & \cdot \\ \cdot & & & \cdot & \cdot \\ 0 & & \cdots & & \bar{r} \end{bmatrix}$$

where $\bar{r} = \lambda_A a_{11}$. ($A = O(\sigma)$.)

(c) If $G \le GL(V)$ then

$$O(G) = \sum_{\sigma \in G} O(\sigma) .$$

A <u>radiation</u> on V is an R-morphism r_α defined for a unit α of R by

$$r_\alpha x = \alpha x$$

for all x in V . Let $RL(V)$ denote the group of radiations of V . It is easy to check that

$$RL(V) = \{\sigma \text{ in } GL(V) \mid \sigma\beta = \beta\sigma \text{ for all } \beta \text{ in } GL(V)\}$$
$$= \text{center of } GL(V) .$$

Further, σ in $GL(V)$ is a radiation if and only if $\sigma L = L$ for all lines L of V. The identity map on V is denoted by 1_V.

The <u>general congruence subgroup of level</u> A (<u>modulo</u> A) for an ideal A is

$$GC(V,A) = \lambda_A^{-1} RL(V/AV) .$$

The kernel of λ_A is denoted by

$$K(V,A) = \{\sigma \text{ in } GL(V) \mid \lambda_A \sigma = 1\} .$$

When dealing with spaces of different dimension the above notation will be subscripted with the dimension. However, we generally suppress the dimension in our notation.

A <u>transvection</u> τ is an element of $GL(V)$ satisfying

(a) there is a hyperplane H of V such that $\tau|_H$ is the identity on H.

(b) for all x in V, $\tau x - x$ is in H.

Finally, the <u>special congruence subgroup of level</u> A (<u>modulo</u> A) for an ideal A of R, denoted $SC(V,A)$, is the subgroup of $GL(V)$ generated by all transvections τ with $0(\tau) \subseteq A$.

A purpose of this chapter is to show that if G is a normal subgroup of $GL(V)$ then there is an ideal A of R such that

$$SC(V,A) \leq G \leq GC(V,A) .$$

Preliminary to this we examine transvections in greater detail.

(B) TRANSVECTIONS

A *transvection* τ is an element of $GL(V)$ satisfying

(a) there is a hyperplane H of V such that $\tau|_H$ is the identity.

(b) for all x in V, $\tau x - x$ is in H.

We call H the *hyperplane* of τ. The identity transvection I_V has many hyperplanes.

A transvection may be approached by way of coördinates by letting $b_1, \ldots, b_{n-1}$ be a basis for H and $b_1, \ldots, b_{n-1}, b_n$ a basis for V. Then

$$\begin{aligned}
\tau b_1 &= b_1 \\
\tau b_2 &= \quad b_2 \\
&\vdots \qquad \ddots \\
\tau b_{n-1} &= \qquad\qquad b_{n-1} \\
\tau b_n &= a_1 b_1 + \cdots + a_{n-1} b_{n-1} + b_n .
\end{aligned}$$

Thus

$$\mathrm{Mat}(\tau) = \begin{bmatrix} 1 & 0 & \cdots & a_1 \\ 0 & 1 & & a_2 \\ \cdot & & \ddots & \vdots \\ \cdot & & & \\ \cdot & & & a_{n-1} \\ 0 & 0 & \cdots & 1 \end{bmatrix} .$$

On the other hand, there is a coordinate-free formulation. Let $\phi : V \to R$ be a surjective linear form with kernel H, $\phi^{-1}(0) = H$.

Select b satisfying $\phi(b) = 1$. This splits $V = H \oplus Rb$. Set $a = \tau b - b$. Then

$$\tau(x) = x + a\phi(x)$$

for all x in V. This follows since $x - b\phi(x)$ lies in the kernel of ϕ, i.e., H. Thus, $\tau x - \tau(b)\phi(x) = x - b\phi(x)$ since $\tau|_H = 1$. Hence $\tau x = x + a\phi(x)$. We say the vector a <u>belongs</u> to τ.

Continuing the above, suppose $\bar{\phi} : V \to R$ is surjective with $\bar{\phi}(H) = 0$. Let $\bar{\phi}(\bar{b}) = 1$ and $\bar{a} = \tau\bar{b} - \bar{b}$. Then

$$\tau x = x + a\phi(x) = x + \bar{a}\bar{\phi}(x) .$$

Thus $a\phi(x) = \bar{a}\bar{\phi}(x)$. Then $a = a\phi(b) = \bar{a}\bar{\phi}(b)$ and $\bar{a} = a\phi(\bar{b})$. Thus $Ra = R\bar{a}$. Since R is local, $a = u\bar{a}$ where u is a unit.[1] Hence a is determined by τ up to a unit multiple.

[1] This follows from (I.2).

If ϕ is replaced by $u\phi$ where u is a unit of R, then there is an $\bar{a}$ in V with

$$\tau(x) = x + a\phi(x) = x + \bar{a}(u\phi)(x) .$$

Thus, since $\phi(b) = 1$, $a = u\bar{a}$. That is, if ϕ is replaced by $u\phi$, u a unit, then a is replaced by $u^{-1}a$. Finally, if a is unimodular, then ϕ is unique up to unit multiple.

We denote the transvection τ in the above discussion by $\tau_{a,\phi}$.

<u>(II.1)</u> <u>LEMMA</u>.

(a) $\tau_{a,\phi} = 1_V$ if and only if $a = 0$.

(b) $\sigma\tau_{a,\phi}\sigma^{-1} = \tau_{\sigma a,\phi\sigma^{-1}}$ for all σ in $GL(V)$.

(c) $\tau_{a,\phi}\tau_{b,\phi} = \tau_{a+b,\phi}$.

(d) If $\tau_{a,\phi} \neq 1$ and a is unimodular then $\tau_{a,\phi} = \tau_{b,\psi}$ if and only if $a = ub$ and $\phi = u^{-1}\psi$ for u a unit.

If τ is a transvection and $\tau \neq 1_V$ we say τ is <u>non-trivial</u>. If $\tau = \tau_{a,\phi}$ where a is unimodular then τ is a <u>unimodular</u> <u>transvection</u>. If $\tau_{a,\phi}$ is unimodular, then Ra is the <u>line</u> of $\tau_{a,\phi}$.

If $b_1,\ldots,b_n$ is a basis for V with dual basis $b_1^*,\ldots,b_n^*$ of V^* then

$$\tau_{\lambda b_i,b_j^*} \qquad (i \neq j)$$

is an elementary transvection.

If $\tau = \tau_{a,\phi}$ is a transvection then $O(\tau) = O(a)$.

(C) THE CONGRUENCE SUBGROUPS OF GL(V)

We assume V is an R-space over a local ring R with dimension ≥ 2 . Recall, if A is an ideal of R then $SC(V,A)$ is the group generated by transvections τ with $O(\tau) \subseteq A$. If $\tau = \tau_{a,\phi}$ since $O(\tau) = O(a)$ and $\sigma\tau\sigma^{-1} = \tau_{\sigma a,\phi\sigma^{-1}}$, then $O(\sigma\tau\sigma^{-1}) = O(\sigma a) = O(a) = O(\tau)$. Thus $SC(V,A)$ is a normal subgroup of $GL(V)$. Further, if $A = R$, by (I.11), $SL(V) = SC(V,R)$ and $GL(V) = SC(V,R)R^*$.

<u>(II.2)</u> <u>THEOREM</u>. Let A be an ideal of R and $\dim(V) \geq 2$. Let $E = \{x \text{ in } V \mid x \text{ is unimodular}\}$. Then $SC(V,A)$ acts as a transformation group on E and there is a natural bijection between the orbits of E and the congruence classes of E modulo A . In particular, $SL(V) = SC(V,R)$ is transitive on E .

<u>Proof</u>. First a special case. Let $\{b_1,\ldots,b_n\}$ be a basis for V and suppose $\alpha = \sum a_i b_i$ is a unimodular vector satisfying $\alpha \equiv b_1 \mod A$. We claim there is a τ in $SC(V,A)$ with $\tau\alpha = b_1$.

Since $a_1 \equiv 1 \mod A$, a_1 is a unit unless $A = R$. In the case $A = R$, since α is unimodular, some a_i is a unit. But, if $A = R$ then $\alpha \equiv b_i \mod A$. Thus, without loss, assume a_1 is a unit. Let E_{ij}

denote the matrix units relative to $\{b_1,\ldots,b_n\}$. Set $\tau_1 = I_V - a_1^{-1}a_2E_{21} - \cdots - a_1^{-1}a_nE_{n1}$. Since $a_2,\ldots,a_n$ are in A , τ_1 is in $SC(V,A)$ and $\tau_1\alpha = a_1b_1$.

If $\sigma = I_V + E_{n1}$, then $\sigma\tau_1\alpha = a_1b_1 + a_1b_n$. Since $a_1 \equiv 1 \mod A$, $a_1 = 1 + s$ for s in A . Set $\tau_2 = I_V - a_1^{-1}sE_{n1}$. Then τ_2 is in $SC(V,A)$ and $\tau_2\sigma\tau_1\alpha = a_1b_1 + b_n$. Letting $\tau_3 = i_V - sE_{n1}$. Then $\tau_3\tau_2\sigma\tau_1\alpha = b_1 + b_n$. Finally, $\sigma^{-1} = I_V - E_{n1}$ gives

$$\sigma^{-1}\tau_3\tau_2\sigma\tau_1\alpha = b_1 .$$

Since τ_3 , τ_2 and τ_1 are in $SC(V,A)$ and $SC(V,A)$ is normal in $GL(V)$,

$$\sigma^{-1}\tau_3\tau_2\sigma\tau_1$$

is in $SC(V,A)$.

Thus we have manufactured a τ in $SC(V,A)$ with $\tau\alpha = b_1$.

Observe the above special case proves the theorem if $A = R$, i.e., $SC(V,R) = SL(V)$ is transitive on E . This fact is used to complete the proof.

Suppose α and β are unimodular and $\alpha \equiv \beta \mod A$. Since β is in E and $SL(V)$ is transitive on E , there is a σ in $SL(V)$ with $\sigma\beta = b_1$. Thus, since $\alpha \equiv \beta \mod A$, $\sigma\alpha \equiv \sigma\beta \mod A$ and hence $\sigma\alpha \equiv b_1 \mod A$. By the first part of the proof there is a τ in

$SC(V,A)$ such that $\tau\sigma\alpha = b_1 = \sigma\beta$, i.e., $(\sigma^{-1}\tau\sigma)\alpha = \beta$. Since $SC(V,A)$ is normal in $SL(V)$, $\sigma^{-1}\tau\sigma$ is in $SC(V,A)$. This completes the proof.

The above result with some changes in the proof is due to Bass. It is a key step in the description of the normal subgroups. Klingenberg utilized a different starting point. This is provided below.

(II.3) THEOREM. Let V be an R-space. Then $GL(V)$ acts transitively on vectors of the same order, i.e., if $O(x) = O(y)$ then there is a σ in $GL(V)$ with $\sigma x = y$.

Proof. Observe, if $\sigma x = y$ for σ in $GL(V)$ then $O(x) = O(y)$.

Suppose $O(x) = O(y) = A$. The last result proved the theorem when $A = R$. Thus assume $A \neq R$.

Let $\{b_1,\ldots,b_n\}$ be a basis for V and suppose $x = \sum a_i b_i$. Then $A = (a_1,\ldots,a_n)$. Let $\{u_1,\ldots,u_p\}$ where $1 \leq p \leq n$ be a minimal generating set for A . Then

$$u_j = \textstyle\sum_i r_{ji}a_i \quad \text{and} \quad a_i = \sum_k s_{ik}u_k .$$

Thus $u_j = \sum\limits_i \sum\limits_k r_{ji}s_{ik}u_k$ and

$$0 = \sum_k (\sum_i r_{ji}s_{ik} - \delta_{jk})u_k \ .$$

Since $\{u_1,\ldots,u_p\}$ is a minimal generating set, $\sum_i r_{ji}s_{ik} - \delta_{jk}$ is not a unit for $1 \leq k \leq p$.

Hence

$$\sum_i r_{ji}s_{ik} - \delta_{jk} \equiv 0 \mod m$$

where m is the maximal ideal of R .

Set $f_t = \sum_i s_{it}b_i$ for $1 \leq t \leq p$. We claim $\{f_t\}$ are linearly independent modulo m . From $0 = \sum c_t f_t = \sum_t c_t (\sum_i s_{it}b_i)$, it follows that $\sum_t c_t s_{it} = 0$. Thus

$$\sum_i (\sum_t s_{it}c_t)r_{ji} = 0 \ .$$

But this is also equal to

$$\sum_i (\sum_t s_{it}c_t)r_{ji} = \sum_t (\sum_i s_{it}r_{ji})c_t$$

$$\equiv c_j \mod m \ .$$

Hence for each j , $c_j \equiv 0 \mod m$.

Thus the $\{f_t\}$ may be completed to a basis of V . Then

$$f_t = \sum s_{jt} b_j .$$

Then
$$x = \sum_i a_i b_i = \sum_i \left(\sum_k s_{ik} u_k\right) b_i$$
$$= \sum_k u_k \left(\sum_i s_{ik} b_i\right)$$
$$= \sum_k u_k f_k .$$

We pause in the proof of (II.3) to single out a corollary of the above proof.

<u>(II.4)</u> <u>PROPOSITION</u>. Let V be an R-space. Let $\{b_1,\ldots,b_n\}$ be a basis of V with $x = \sum a_i b_i$ and $A = (a_1,\ldots,a_n)$. Then if $\{u_1,\ldots,u_p\}$ is a minimal generating set for A $(p \le n)$, there is a basis $\{f_1,\ldots,f_n\}$ of V with $x = u_1 f_1 + \cdots + u_p f_p$.

To complete the proof of (II.3), apply the above also to y , $O(y) = A$. Thus, there is a basis $\{g_i\}$ of V with $y = \sum u_i g_i$. Define $\sigma : V \to V$ by $\sigma f_i = g_i$. Then σ is in $GL(V)$ and $\sigma x = y$.

<u>(II.5)</u> <u>THEOREM</u>. Let V be an R-space with $\dim(V) \ge 2$. Let $SL(V)$ act on itself as a transformation group by conjugation, i.e., if σ is in $SL(V)$ define $\rho_\sigma : SL(V) \to SL(V)$ by $\rho_\sigma(\beta) = \sigma\beta\sigma^{-1}$. Then the orbit of a transvection τ consists of all transvections of the same order as τ .

Proof. Clearly transvections are carried by conjugation into transvections of the same order. Suppose conversely $\tau_1 = \tau_{a_1,\phi_1}$ and $\tau_2 = \tau_{a_2,\phi_2}$ are transvections of the same order, i.e., $0(a_1) = 0(a_2)$. Let $H_1 = \phi_1^{-1}(0)$ and $H_2 = \phi_2^{-1}(0)$ be the corresponding hyperplanes where

$$H_1 = \oplus \sum_{i=1}^{n-1} Re_i$$

and

$$H_2 = \oplus \sum_{i=1}^{n-1} Rf_i \ .$$

Then $a_1 = \alpha_1 e_1 + \cdots + \alpha_{n-1} e_{n-1}$, $a_2 = \beta_1 f_1 + \cdots + \beta_{n-1} f_{n-1}$, and $0(a_1) = (\alpha_1,\ldots,\alpha_{n-1}) = (\beta_1,\ldots,\beta_{n-1}) = 0(a_2)$. Let $\{e_1,\ldots,e_n\}$, $\{f_1,\ldots,f_n\}$ be bases of V.

To manufacture a σ with $\sigma\tau_1\sigma^{-1} = \tau_2$, Klingenberg began by finding a σ in $GL_{n-1}(V)$ with $\sigma a_1 = a_2$ by (II.3). On the other hand, Bass added the additional basis vectors e_n and f_n to a_1 and a_2, respectively. That is, $a_1 + e_n$ and $a_2 + f_n$ which are unimodular and essentially congruent modulo A, and then utilized (II.2).

Create a suitable σ as follows: (using (II.2))

$$
\begin{array}{l}
a_1 + e_n \\
\quad \Big\downarrow \tau \qquad (\tau \text{ in } SC(V,A)\ ,\ \tau e_n = e_n\ ,\ \tau(H_1) \subseteq H_1) \\
e_n \\
\quad \Big\downarrow \beta \qquad (\beta \text{ in } SL(V)\ ,\ \beta e_n = f_n\ ,\ \beta H_1 = H_2) \\
f_n \\
\quad \Big\downarrow \tau' \qquad (\tau' \text{ in } SC(V,A)\ ,\ \tau' f_n = f_n\ ,\ \tau'(H_2) \subseteq H_2) \\
a_2 + f_n
\end{array} .
$$

Then $a_2 + f_n = \sigma(a_1 + e_n) = \sigma(a_1) + \sigma(e_n) = \sigma(a_1) + f_n$ where $\sigma = \tau'\beta\tau$. Then $\sigma(a_1) = a_2$. Consider $\sigma\tau_1\sigma^{-1} = \tau_{\sigma a_1, \phi_1\sigma^{-1}}$. Since $\sigma a_1 = a_2$ and $(\phi_1\sigma^{-1})^{-1}(0) = \sigma\phi_1^{-1}(0) = \sigma H_1 = H_2$, we have $\tau_{\sigma a_1, \phi_1\sigma^{-1}} = \tau_{a_2, \phi_2}$ as desired.

Let G be a group and a and b elements of G. The element $[a,b] = a^{-1}b^{-1}ab$ is the <u>commutator</u> of a and b. If H and K are subgroups of G then $[H,K]$ denotes the subgroup generated by all $h^{-1}k^{-1}hk$ where h is in H and k in K and is called a <u>mixed commutator subgroup</u> of G determined by H and K. The subgroup $[G,G]$ is the <u>commutator subgroup</u> of G.

<u>(II.6)</u> <u>THEOREM</u>. (Characterizations of $SC(V,A)$) Let A be an ideal of R. Let V be an R-space of dimension ≥ 2. Then, the following subgroups of $GL(V)$ coincide:

(a) The group $SC(V,A)$ generated by transvections τ with

$0(\tau) \subseteq A$.

(b) The group $H_A = \{\sigma \text{ in } GL(V) \mid \det(\sigma) = 1 \text{ and } \lambda_A\sigma = I\}$ where $\lambda_A : GL(V) \to GL(V/AV)$.

If, in addition, $\dim(V) \geq 3$ the above are the same as

(c) $K_A = [GL(V),GC(V,A)]$.[1]

Proof. Let $G_A = SC(V,A)$. The proof is in several steps.

(a) Note $G_A \subseteq H_A$ since G_A is generated by transvections τ with $0(\tau) \subseteq A$. Hence $\det(\tau) = 1$ and $\lambda_A\tau = I$.

(b) To show $H_A \subseteq G_A$, let σ be in H_A . Thus $\det(\sigma) = 1$ and $\lambda_A(\sigma) = I$. Relative to a basis of V ,

$$\text{Mat}(\sigma) = \begin{bmatrix} 1+a_{11} & a_{12} & \cdots & a_{1n} \\ a_{21} & 1+a_{22} & \cdots & a_{2n} \\ \vdots & \vdots & & \vdots \\ a_{n1} & a_{n2} & \cdots & 1+a_{nn} \end{bmatrix}$$

where a_{ij} are in A . The first column is congruent to $\langle 1,0,\ldots,0\rangle^t$ modulo A .[2] By (II.2), select a suitable τ_1 element of $SC(V,A)$ which carries the first column to $\langle 1,0,\ldots,0\rangle^t$. Applying τ_1 to $\text{Mat}(\sigma)$ sweeps out the first column, leaving $\langle 1,0,\ldots,0\rangle^t$. The congruence class of the second column of $\text{Mat}(\sigma)$ is the same as the second column of $\tau_1\text{Mat}(\sigma)$. Continue this process. By successive applications of elements τ_i in $SC(V,A)$ we obtain

---

[1] It is easy to see $GC(V,A)$ is the maximal subgroup G of $GL(V)$ satisfying $SC(V,A) = [GL(V),G]$.

[2] t denotes the transpose.

$$\tau_n \tau_{n-1} \cdots \tau_1 \mathrm{Mat}(\sigma) = I .$$

Hence $\mathrm{Mat}(\sigma) = \tau_1^{-1}\tau_2^{-1}\cdots\tau_n^{-1}$. That is, σ is in $SC(V,A)$, i.e., $H_A \subseteq G_A$.

(c) Observe $K_A \subseteq H_A$ since a generator of K_A has the form $\sigma^{-1}\beta^{-1}\sigma\beta$ where σ is in $GL(V)$ and β is in $GC(V,A)$. Thus $\det(\sigma^{-1}\beta^{-1}\sigma\beta) = 1$ and $\lambda_A(\sigma^{-1}\beta^{-1}\sigma\beta) = \lambda_A(\sigma^{-1})\lambda_A(\sigma)\lambda_A(\beta^{-1})\lambda_A(\beta) = I$.

(d) It remains to show $H_A \subseteq K_A$. Assume $\dim(V) \geq 3$. Then if H is a hyperplane, $\dim(H) \geq 2$.

Let $\tau = \tau_{a,\phi}$ be a transvection in $H_A = SC(V,A)$ with $H = \phi^{-1}(0)$. Then a is in H . We show $a = a_2 - a_1$ where a_2 and a_1 are in H and $0(a) = 0(a_1) = 0(a_2)$.

Let $a = \sum_{i=1}^{n} \alpha_i b_i$ where $\{b_i\}_{i=1}^{n-1}$ is a basis for H .

<u>Case (1)</u>. $n - 1 = 2m$.

Set
$$a_1 = \sum (\alpha_{2j} - \alpha_{2j-1})b_{2j-1} + \sum \alpha_{2j-1}b_{2j} ,$$
$$a_2 = \sum \alpha_{2j}b_{2j-1} + \sum (\alpha_{2j-1} + \alpha_{2j})b_{2j} .$$
Then $a_2 - a_1 = a$ and $0(a) = 0(a_1) = 0(a_2)$.

<u>Case (2)</u>. $n - 1 = 2m + 1$.

Use the above formula except replace the $j = 1$ summand by

$$(\alpha_{2m+1} + \alpha_2 - \alpha_1)b_1 + (\alpha_1 + \alpha_{2m+1})b_2 + \alpha_1 b_{2m+1}$$

in the expression for a_1 and by

$$(\alpha_2 + \alpha_{2m+1})b_1 + (\alpha_1 + \alpha_2 + \alpha_{2m+1})b_2 + (\alpha_{2m+1} + \alpha_1)b_{2m+1}$$

in the expression for a_2. Again, it is easy to check that $a_2 - a_1 = a$ and $O(a) = O(a_1) = O(a_2)$.

Define $\tau_1 = \tau_{a_1,\phi}$ and $\tau_2 = \tau_{a_2,\phi}$. Since $O(a_1) = O(a_2)$ there is a σ in $GL(V)$ with $\tau_2 = \sigma\tau_1\sigma^{-1}$. Then

$$\tau_2\tau_1^{-1} = \sigma\tau_1\sigma^{-1}\tau_1^{-1}$$

is in K_A, but

$$\tau_2\tau_1^{-1} = \tau_{a_2,\phi}\tau_{a_1,\phi}^{-1} = \tau_{a_2,\phi}\tau_{-a_1,\phi}$$

$$= \tau_{a_2-a_1,\phi} = \tau_{a,\phi} = \tau .$$

Hence $H_A \subseteq K_A$.

<u>(II.7)</u> <u>COROLLARY</u>. Let A be an ideal of R and V be an R-space of dimension ≥ 2. Then,

(a) $GC(V,A)/SC(V,A)$ is isomorphic to the subgroup of the group $R^* \times (R/A)^*$ consisting of all pairs $\langle a,b \rangle$ with $\Pi_A a = b^n$.

(b) If $HC(V,A) = GC(V,A) \cap SL(V)$, then $HC(V,A)/SC(V,A)$ is isomorphic to the group of n-th roots of unity in $(R/A)^*$.

Proof. Define $\phi : GC(V,A) \to R^* \times (R/A)^*$ by $\phi(\sigma) = \langle\det(\sigma),\lambda_A(\sigma)\rangle$. Then $\mathrm{Ker}(\phi) = SC(V,A)$. Let σ be in $GC(V,A)$ with $\det(\sigma) = a$. Then $\Pi_A(a) = \Pi_A(\det \sigma) = \det(\lambda_A\sigma) = \det(bI) = b^n$. To show ϕ is onto select $\langle a,b\rangle$ in $R^* \times (R/A)^*$ with a in R^* , b in $(R/A)^*$ and $\Pi_A a = b^n$. Select b' in R^* with $\Pi_A b' = b$. Form

$$\sigma = \begin{bmatrix} b'+y & 0 & \cdots & 0 \\ 0 & b' & & \cdot \\ \vdots & & \ddots & \vdots \\ 0 & 0 & \cdots & b' \end{bmatrix}$$

where y is to be determined. We want $\det(\sigma) = a$, but $\det(\sigma) = b'^n + b'^{n-1}y$. Since $b'^n \equiv a \mod A$, i.e., $a = b'^n + z$ for z in A , take $y = (b'^{n-1})^{-1}z$. This choice gives $\det(\sigma) = a$ and $\phi : \sigma \to \langle\det(\sigma),\lambda_A(\sigma)\rangle = \langle a,b\rangle$.

(D) THE NORMAL SUBGROUPS OF GL(V)

Since (II.7) indicates $GC(V,A)/SC(V,A)$ is an Abelian group, if H is a subgroup of $GL(V)$ with $SC(V,A) \le H \le GC(V,A)$ then H is normal in $GC(V,A)$ and, in turn, normal in $GL(V)$. The purpose of this section is to show the converse is true. That is, if H is a normal subgroup of $GL(V)$ then there is an ideal A of R with $SC(V,A) \le H \le GC(V,A)$.

The method of proof is to first show that if H is a normal subgroup of $GL(V)$ then H contains a transvection. Next, it is shown that if τ is a transvection with $0(\tau) = A$ then

$$\{\sigma\tau\sigma^{-1} \mid \sigma \text{ in } SL(V)\}$$

is precisely $SC(V,A)$. More precisely, this is done for a family of transvections whose orders generate A. Thus, applying this to the normal subgroup H we create an ideal A with $SC(V,A) \leq H$. Finally, it is shown that $H \leq GC(V,A)$.

(II.8) THEOREM. Let V be an R-space of dimension ≥ 3. Let $\{\tau_\alpha\}_{\alpha\in\Lambda}$ be a set of transvections with $0(\tau_\alpha) = J_\alpha$. Let $J = \sum_\alpha J_\alpha$. Then the SL(V)-normal[1] subgroup of $GL(V)$ generated by $\{\tau_\alpha\}$ is $SC(V,J)$.

Proof. We assume initially that $\{\tau_\alpha\}_{\alpha\in\Lambda} = \{\tau\}$. Let $\tau = \tau_{a,\phi}$ where $a = \alpha_1 e_1 + \cdots + \alpha_{n-1}e_{n-1}$, $H = \phi^{-1}(0) = \sum_{i=1}^{n-1} Re_i$ and $J = 0(\tau) = 0(a) = (\alpha_1,\ldots,\alpha_{n-1})$.

Define

$$a_1 = \sum_{i=1}^{n-1} \alpha_i e_i - \sum_{i=2}^{n-1} \alpha_{i-1}e_i$$

$$= \alpha_1 e_1 + (\alpha_2 - \alpha_1)e_2 + \cdots + (\alpha_{n-1} - \alpha_{n-2})e_{n-1}$$

and

$$a_2 = \sum_{i=2}^{n-1} \alpha_{i-1}e_i .$$

[1] This means a subgroup of $GL(V)$ normalized by $SL(V)$.

Thus $O(a_1) = O(a)$ and $a = a_1 + a_2$. Let $\tau_1 = \tau_{a_1,\phi}$ and $\tau_2 = \tau_{a_2,\phi}$. Then $\tau_{a,\phi} = \tau_1\tau_2$. By (II.5) τ_1 lies in the orbit of τ , that is, there exists a σ in SL(V) with $\sigma\tau\sigma^{-1} = \tau_1$.

Thus, if G denotes the SL(V)-normal subgroup generated by τ , then τ_1 is in G . Thus, $\tau_2 = \tau_1^{-1}\tau$ is in G . But $\tau_2 = \tau_{a_2,\phi}$ where

$$a_2 = \alpha_1 e_2 + \cdots + \alpha_{n-2}e_{n-1}$$

has $n - 2$ summands. Continuing, one finds that G contains a transvection having associated vector $\alpha_1 e_{n-1}$. Again, by (II.5), G contains all transvections of order (α_1) . By reindexing $\{e_1,\ldots,e_{n-1}\}$ and applying the above argument, deduce that G contains all transvections with order (α_i) for $1 \le i \le n - 1$.

Let τ' be a transvection with $O(\tau') \subset J = O(\tau)$. Then $\tau' = \tau'_{a',\phi'}$ where $a' = \sum \alpha_i' e_i'$, $H' = (\phi')^{-1} = \sum Re_i'$ and

$$(\alpha_1',\ldots,\alpha_{n-1}') \subset J = (\alpha_1,\ldots,\alpha_{n-1}) .$$

Thus, for each i , $1 \le i \le n - 1$,

$$\alpha_i' = \sum_j \alpha_{ij}\alpha_j .$$

By a straightforward construction G contains all transvections of order (α_i') . Thus $\tau_i' = \tau_{\alpha_i' e_i',\phi'}$ is in G for each i and

$$\tau' = \prod_i \tau_i'$$

belongs to G .

We now consider an arbitrary family $\{\tau_\alpha \mid \alpha \text{ in } \Lambda\}$ of transvections. Let $J = \sum J_\alpha$ and $J_\alpha = O(\tau_\alpha)$. Suppose τ' is an arbitrary transvection with $O(\tau') = J' \subset J$ where $\tau' = \tau_{a',\phi'}$ and $a' = \sum \alpha_i' e_i'$. The above argument provides transvections τ_i', $1 \le i \le n - 1$, of the form $\tau_i'(x) = x + \alpha_i' e_i' \phi'(x)$. Further, $\tau' = \Pi\, \tau_i'$.

Theorem (II.8) permits the completion of the normal subgroup problem.

(II.9) THEOREM. (Normal Subgroups of $GL(V)$)[1] Let V be an R-space with $\dim(V) \ge 3$. Let G be a subgroup of $GL(V)$ with $O(G) = A$. Then G is $SL(V)$-normal if and only if $SC(V,A) \le G \le GC(V,A)$.

Proof. (Bass) Suppose G is $SL(V)$-normal. The proof that $SC(V,A) \le G \le GC(V,A)$ is in several steps. This is the difficult case.

(1) We show the centralizer of $SL(V)$ consists of all scalar matrices.

(2) If $n \ge 3$ and G contains a non-central element with some coordinate 0, then we show G contains transvections of order contained in A.

(3) If G contains a non-central element, we produce an element in G with a coordinate 0.

To show (1), consider a fixed basis for V and relative to this basis

$$T = \{\tau = I + aE_{12} \mid a \text{ in } R\}.$$

---

[1] For the case $\dim(V) = 2$ see the Appendix to (IV).

We want to compute the centralizer

$$C(T) = \{\sigma \text{ in } GL(V) \mid \sigma\tau = \tau\sigma \text{ for all } \tau \text{ in } T\}$$

of T. We use extensively in the proof the following notation. Let σ be in $C(T)$ and

$$\alpha = \text{first column of } \sigma$$
$$\beta = \text{second row of } \sigma^{-1} .$$

Then for τ in T,

$$\begin{aligned} \sigma\tau\sigma^{-1} &= \sigma(I + aE_{12})\sigma^{-1} \\ &= I + \sigma(aE_{12})\sigma^{-1} \\ &= I + \alpha a\beta . \end{aligned}$$

But $\sigma\tau\sigma^{-1} = \tau$, therefore $aE_{12} = \alpha a\beta$ for all a in R. In particular, if $a = 1$ then $E_{12} = \alpha\beta$ and

$$\alpha = \langle u,0,\ldots,0\rangle^t \qquad ({}^t \text{ denotes transpose})$$
$$\beta = \langle 0,u^{-1},0,\ldots,0\rangle$$

where u is a unit in R. Applying this discussion to all E_{ij}, we find the centralizer of $SL(V)$ consists of scalar matrices.

We continue the notation of the above paragraph. Assume that σ is a non-central element in G with a coordinate equal to 0. Since a sufficient quantity of permutation matrices lie in $SL(V)$, by conjugating by a suitable permutation, assume σ has first column

$$\alpha = \langle a_1, \ldots, a_n \rangle^t$$

where either $a_1 = 0$ or $a_n = 0$ (remember G is SL(V)-normal).

Two cases:

(a) Suppose σ commutes with all $\tau = I + aE_{12}$. By the above discussion

$$\sigma = \begin{bmatrix} u & \\ 0 & \\ \vdots & * \\ 0 & \end{bmatrix}.$$

By conjugating by permutation matrices, assume

$$\sigma = \begin{bmatrix} & 0 \\ & \vdots \\ * & \\ & 0 \\ & u \end{bmatrix}.$$

Then $\sigma = u\sigma'$ where

$$\sigma' = \left[\begin{array}{c:c} & 0 \\ \bar{\sigma}' & \vdots \\ & 0 \\ \hdashline \delta & 1 \end{array}\right].$$

Recall the affine group Aff(n - 1,R) is the semi-direct product of $GL_{n-1}(R)$ and R^{n-1}, i.e.,

$$Aff(n - 1,R) = GL_{n-1}(R) \times R^{n-1} ,$$

where if $\langle\beta,\alpha\rangle$ and $\langle\beta',\alpha'\rangle$ are in Aff(n - 1,R), then multiplication is given by

$$\langle\beta,\alpha\rangle\langle\beta',\alpha'\rangle = \langle\beta\beta',\alpha\beta' + \alpha'\rangle$$

and inverse

$$\langle\beta,\alpha\rangle^{-1} = \langle\beta^{-1},-\alpha\beta^{-1}\rangle .$$

The above constructions permit us to identify σ' with an element $\langle\bar{\sigma}',\delta\rangle$ of Aff(n - 1,R). Similarly, Aff(n - 1,R) may be embedded in $GL_n(R)$ by employing matrices of the form of σ'.

Viewing σ' as an element of Aff(n - 1,R), we employ a basic commutator relation which produces the desired transvection

$$[\langle\bar{\sigma}',\delta\rangle,\langle I,\tau\rangle] = \langle\bar{\sigma}',\delta\rangle^{-1}\underbrace{\langle I,\tau\rangle^{-1}\langle\bar{\sigma}',\delta\rangle\langle I,\tau\rangle}_{\text{in } G}$$

$$= \langle I,\tau(I - \bar{\sigma}')\rangle .$$

Observe u plays no role in the commutator. Hence we obtain a transvection $\langle I,\tau(I - \bar{\sigma}')\rangle$ in G.

(b) Suppose there is a $\tau = I + aE_{12}$ with $\rho = \sigma\tau\sigma^{-1}\tau^{-1} \neq I$. Observe, with the above notation, $\sigma\tau\sigma^{-1} = I + \alpha a\beta$. Therefore

$$\rho = \tau^{-1} + \alpha\delta$$

where $\delta = a\beta\tau^{-1}$.

Let $\alpha = \langle a_1,\ldots,a_n\rangle^t$.

Two subcases occur:

(1) In α , $a_n = 0$ and thus the last row of $\alpha\delta$ is zero.

(2) In α , $a_1 = 0$ and thus the first row of $\alpha\delta$ is zero. If this case occurs, the first row of ρ (which is the first row of τ^{-1}) is $\langle 1,-a,0,\ldots,0\rangle$. Thus ρ is non-central and, since $n \geq 3$, has a zero in the first row. By suitable permutation, place the zero in (n,1)-position. This returns us to Case (1).

Assume $a_n = 0$. Then the last row of $\alpha\delta$ is zero and

$$\rho = \begin{bmatrix} & * & \\ & & \\ 0,\ldots,0,1 \end{bmatrix} .$$

By transposing, the argument in (a) utilizing the affine group may be repeated to produce a transvection.

We now show (3), that if G contains a non-central element then it is possible to produce by conjugation by elements in $SL(V)$ an element with a position equal to 0.

Let σ be non-central. Let

$$\alpha = \langle a_1, \ldots, a_n \rangle^t$$

with $(a_1, \ldots, a_n) = R$ be the first column of σ. Conjugation by a suitable transvection will place a unit in the (1,1)-position, thus assume a_1 is a unit. Select a transvection λ so that

$$\lambda^{-1}\alpha = \langle *, \ldots, *, 0 \rangle^t .$$

But, (warning) $\lambda^{-1}\sigma$ may not be in G.

If σ commutes with all $\tau = I + aE_{12}$, then the first column of σ has zero entries and we are done. Hence assume there is a τ with $\rho = \sigma\tau\sigma^{-1}\tau^{-1} \neq I$. As before, $\rho = \tau^{-1} + \alpha\delta$ where $\delta = a\beta\tau^{-1}$.

We claim that ρ is not central. Suppose ρ is central. Then $\rho = vI$ for some v a unit. With λ chosen as above the last row of $(\lambda^{-1}\alpha)\delta$ is zero. Hence $\lambda^{-1}\rho = \lambda^{-1}\tau^{-1} + \lambda^{-1}\alpha\delta$ and the last row of $\lambda^{-1}\rho$ is the same as the last row of $\lambda^{-1}\tau^{-1}$. Since $n \geq 3$, $\lambda^{-1}\tau^{-1}$ has 1 in (n,n)-position and hence $\lambda^{-1}\rho$ has 1 in (n,n)-position. But if ρ is central, $\lambda^{-1}\rho\lambda = \lambda^{-1}\lambda\rho = \rho$ and this has 1 in (n,n)-position. Hence $v = 1$ and $\rho = I$ — a contradiction.

Therefore ρ is not central. Then, $\lambda^{-1}\rho\lambda$ is not central, is contained in G and has zeros in the last row except for possibly two positions. But $n \geq 3$ — done.

The above proof shows that if $G \leq GL(V)$ is normalized by $SL(V)$ and G is not in the center then G contains transvections. Let $\{\tau_\alpha\}$ with $O(\tau_\alpha) = J_\alpha$ be transvections in G. Then $SC(V,J) \leq G$ where $J = \sum J_\alpha$.

Select J maximal with respect to $SC(V,J) \leq G$. We claim that under the map

$$\lambda_J : GL(V) \to GL(V/JV)$$

the image $\bar{G} = \lambda_J(G)$ lies in the center of $GL(V/JV)$. This follows since if $SL(V)$ normalizes G then $SL(V/JV)$ normalizes $\bar{G}$. If $\bar{G}$ is non-central, since R/J is local, repeat the previous arguments and find $\bar{J}$ satisfying $SC(V/JV,\bar{J}) \leq \bar{G}$. By the Correspondence Theorem there is an ideal J^* of R, properly containing J, with $\bar{J} = J^*/J$. Then the inverse image of $\bar{G}$ contains $SC(V,J^*)$. This contradicts the maximality of J. Hence $\lambda_J G \leq \text{Center}(GL(V/JV))$. On the other hand, $O(G) = A$ is the smallest ideal of R satisfying $\lambda_A G \subseteq \text{Center}(GL(V/AV))$. It is easy to see that $A = J$.

Suppose, conversely, that G is a subgroup of $GL(V)$ with $SC(V,A) \leq G \leq GC(V,A)$. If σ is in $GL(V)$ and τ is in G then $\sigma\tau\sigma^{-1}\tau^{-1}$ is in $SC(V,A)$ by (II.6) and hence in G. Thus $\sigma\tau\sigma^{-1}$ is in G and G is normal.

(E) INVOLUTIONS

Let R denote a local ring with 2 a unit, maximal ideal m and residue field $k = R/m$. Let V be an R-space with dimension ≥ 3.

The previous sections concerned the classification of the normal subgroups of $GL(V)$. We examine in the remaining sections the description of the group automorphisms of $GL(V)$. The approach is by the 'method of involutions' developed by Dieudonné [63] and Rickart [119]. These sections follow the work of Keenan [88]. A discussion of alternate approaches occurs in the concluding section of this chapter.

We describe three 'simple' automorphisms of $GL(V)$: (1) those automorphisms P_χ of $GL(V)$ which can be expressed in the form $P_\chi(\sigma) = \chi(\sigma)\sigma$ where χ is a group morphism of $GL(V)$ into its center; (2) those automorphisms ϕ_g of $GL(V)$ which can be expressed in the form $\phi_g(\sigma) = g\sigma g^{-1}$, for some semi-linear automorphism $g : V \to V$; (3) those automorphisms ψ_g of $GL(V)$ which can be expressed in the form $\psi_g(\sigma) = g^{-1}\check{\sigma}g$, for some semi-linear isomorphism $g : V \to V^*$ where V^* is the dual of V.[1] The main result will then become: Every automorphism Λ of $GL(V)$ can be expressed uniquely as either

$$\Lambda = P_\chi \circ \phi_g \qquad \text{or} \qquad \Lambda = P_\chi \circ \psi_g .$$

To show Λ has the above form, begin by noting that if $\sigma^2 = 1$ where

---

[1] The definition of $\check{\sigma}$ is given in (II.F).

σ is in $GL(V)$ then $[\Lambda(\sigma)]^2 = 1$. An element σ with $\sigma^2 = 1$ is called an _involution_. Hence an automorphism Λ carries involutions to involutions. We will employ this observation together with properties of involutions, to show Λ induces naturally a projectivity on V . Invoking the Fundamental Theorem of Projective Geometry produces a semi-linear isomorphism g from which we create the desired form of Λ .

This section will collect the necessary results on involutions.

Let σ be in $GL(V)$ with $\sigma^2 = 1$. Then σ is called an _involution_. Each involution σ determines a unique splitting $V = P(\sigma) \oplus N(\sigma)$ where $\sigma = 1_{P(\sigma)} \oplus -1_{N(\sigma)}$; precisely,

$$P(\sigma) = \{x \text{ in } V \mid \sigma(x) = x\} ,$$

$$N(\sigma) = \{x \text{ in } V \mid \sigma(x) = -x\} .$$

Call $P(\sigma)$ the _positive_ space of σ and $N(\sigma)$ the _negative_ space of σ . Observe, it is important to have 2 a unit to show $N(\sigma) \cap P(\sigma) = 0$ and $V = N(\sigma) + P(\sigma)$ (note $x = \frac{1}{2}(x + \sigma x) + \frac{1}{2}(x - \sigma x)$). The spaces $N(\sigma)$ and $P(\sigma)$ are the _proper_ spaces of σ and vectors in these spaces are the _proper_ vectors of σ .

An involution σ is _extremal_ if either $P(\sigma)$ is a line or $P(\sigma)$ is a hyperplane, i.e., $\dim(P(\sigma)) = 1$ or $\dim(P(\sigma)) = n - 1$.

(II.10) LEMMA.

(a) If σ_1 and σ_2 are involutions, then $\sigma_1 = \pm\sigma_2$ if and only if σ_1 and σ_2 have the same proper spaces.

(b) If σ is an involution and ρ is in $GL(V)$ then $\rho\sigma\rho^{-1}$ is an involution with proper spaces $\rho N(\sigma)$ and $\rho P(\sigma)$. In particular, $\rho\sigma = \sigma\rho$ if and only if $\rho N(\sigma) = N(\sigma)$ and $\rho P(\sigma) = P(\sigma)$. (A similar statement is true if ρ is invertible semi-linear.)

If σ_1 and σ_2 are involutions, then $\sigma_1\sigma_2$ is an involution if and only if σ_1 and σ_2 commute. Thus, any group of involutions must be commutative. If X is any set of pairwise commuting involutions, then there exist lines $L_1,\ldots,L_n$ with

$$V = L_1 \oplus \cdots \oplus L_n$$

and $\sigma L_i = L_i$ for $1 \le i \le n$ and σ in X. Thus, each σ in X will have the form

$$\sigma = (\pm 1_{L_1}) \oplus \cdots \oplus (\pm 1_{L_n}) .$$

If X is a group, then X is commutative and $|X| \le 2^n$.

Let σ be an involution with $p = \dim(P(\sigma))$ and $m = \dim(N(\sigma))$. Then σ is said to be a (p,m)-*involution* or of *type* (p,m).

(II.11) LEMMA.

(a) There exist at most $\binom{n}{p}$ elements in any collection of pairwise commuting involutions of type $(p,n-p)$.

(b) In any set of pairwise commuting involutions there exist at most 2^n elements.

(c) $GL(V)$ is transitive under conjugation on the sets of involutions of the same type.

Denote by Ext the set of <u>all</u> extremal involutions in $GL(V)$, i.e., the set of all involutions σ with $\dim(P(\sigma)) = 1$ or $\dim(P(\sigma)) = n - 1$.

<u>(II.12)</u> <u>THEOREM</u>. Let Λ be an automorphism of $GL(V)$. Then $\Lambda(\mathrm{Ext}) = \mathrm{Ext}$, i.e., the set of extremal involutions will be carried onto itself under the action of Λ .

<u>Proof</u>. Let σ be an extremal involution. Then σ may be extended to a maximal set X of 2^n mutually commuting involutions. Further, by (II.11), σ has $n = \binom{n}{1}$ conjugates in X . Then $\Lambda(X)$ is a set of 2^n mutually commuting involutions and $\Lambda\sigma$ has exactly $\binom{n}{1}$ conjugates in $\Lambda(X)$. (Note: if $1 < p < n - 1$ then $\binom{n}{1} < \binom{n}{p}$.) Thus, $\Lambda\sigma$ has type $(1,n - 1)$ or $(n - 1,1)$ and hence is extremal.

In the above proof we used the facts that Λ is bijective, preserves involutions, conjugation and pairwise permutability. In fact, the observation that an involution of type $(p,n - p)$ has only $\binom{n}{p}$ conjugates in a maximal system of pairwise commuting involutions is purely group theoretic. It is the conversion of geometric algebra concepts into group theoretic concepts that will allow the description of Λ to be obtained. A similar undercurrent was present in the classification of the normal subgroups.

Let X and Y be subsets of $GL(V)$ and Λ an automorphism of $GL(V)$. The _centralizer_ of Y _by_ X is

$$C_X(Y) = \{\sigma \text{ in } X \mid \sigma\beta = \beta\sigma \text{ for all } \beta \text{ in } Y\}.$$

Observe $\Lambda C_X(Y) = C_{\Lambda X}(\Lambda Y)$.

_(II.13)_ _COROLLARY._ Let Λ be an automorphism of $GL(V)$. Then

$$\Lambda C_{Ext}(Y) = C_{Ext}(\Lambda Y).$$

for Y a subset of $GL(V)$.

Suppose σ is an extremal involution. Let

$$L(\sigma) = \text{proper line of } \sigma$$

$$H(\sigma) = \text{proper hyperplane of } \sigma.$$

Observe $L(\sigma)$ and $H(\sigma)$ determine σ up to a unit factor.

We now begin to utilize extensively the remarks after (II.10). To repeat, suppose involutions σ and β commute. Then V splits $V = L_1 \oplus \cdots \oplus L_n$ where the proper spaces of σ and β are direct sums of subsets of the lines $L_1, \ldots, L_n$.

On the other hand, we wish to borrow the standard theory of involutions from V/mV . Thus, if σ and β do not commute then their images should not commute in $GL(V/mV)$.

Recall Π_m denotes the morphism $V \to V/mV$ and λ_m the morphism $GL(V) \to GL(V/mV)$.

Let σ and β be extremal involutions. The pair $<\sigma,\beta>$ is a _minimal couple_ if

(a) $\lambda_m\sigma\beta \neq \lambda_m\beta\sigma$

(b) Either $L(\sigma) = L(\beta)$ or $H(\sigma) = H(\beta)$.

To illustrate the above concept we treat two cases.

(I) $<\sigma,\beta>$ is a minimal couple and $L = L(\sigma) = L(\beta)$. Then $\Pi_m H(\sigma) \neq \Pi_m H(\beta)$ for otherwise $\lambda_m\sigma\beta$ would equal $\lambda_m\beta\sigma$. By (I.7) - _an important lemma_ - $W = H(\sigma) \cap H(\beta)$ is a subspace of dimension $n - 2$ in V . Hence, if x is in $L \oplus W$, $\sigma x = \pm x$ and $\beta x = \pm x$. Pictorially,

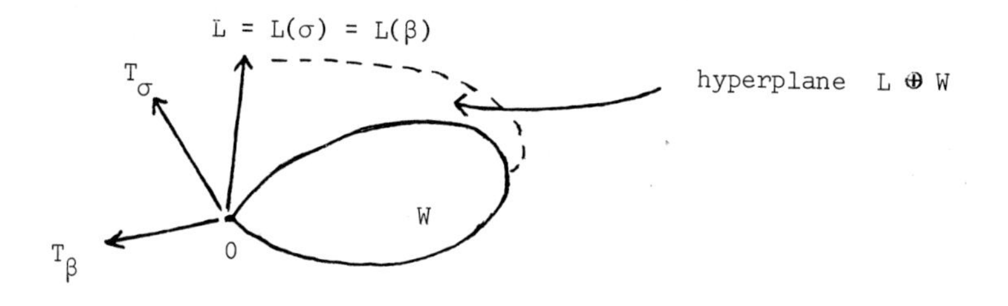

T_σ and T_β represent the "extra" lines of σ and β belonging to their respective hyperplanes.

(II) $<\sigma,\beta>$ is a minimal couple and $H = H(\sigma) = H(\beta)$. Pictorially,

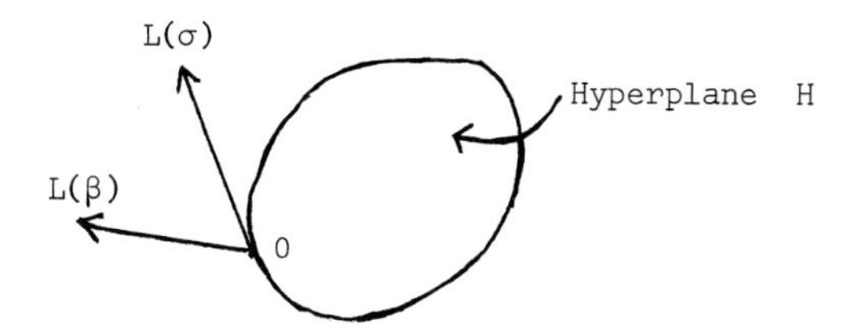

Suppose $<\sigma,\beta>$ is a minimal couple of extremal involutions. Consider the centralizer of $\{\sigma,\beta\}$ by extremal involutions:

$$C = C_{\mathrm{Ext}}(\{\sigma,\beta\}) .$$

Assume $L = L(\sigma) = L(\beta)$. If α is in C , i.e., $\alpha\sigma = \sigma\alpha$ and $\alpha\beta = \beta\alpha$, then $L(\alpha)$ must be contained in $H(\sigma) \cap H(\beta)$ (for if in a splitting of V we had $L(\alpha) = L(\beta)$ then $\alpha = \pm\beta$ and thus β would commute with σ - a contradiction). Hence L is in $H(\alpha)$. Conversely, any α in Ext with $L(\alpha) \subset H(\sigma) \cap H(\beta)$ and $L \subset H(\alpha)$ is in C . This follows since under this assumption, say $L(\alpha) \subset H(\sigma)$ and $L \subset H(\alpha)$, then $\Pi_m H(\sigma) \neq \Pi_m H(\alpha)$. Let $W = H(\sigma) \cap H(\alpha)$ and apply (I.7). We have

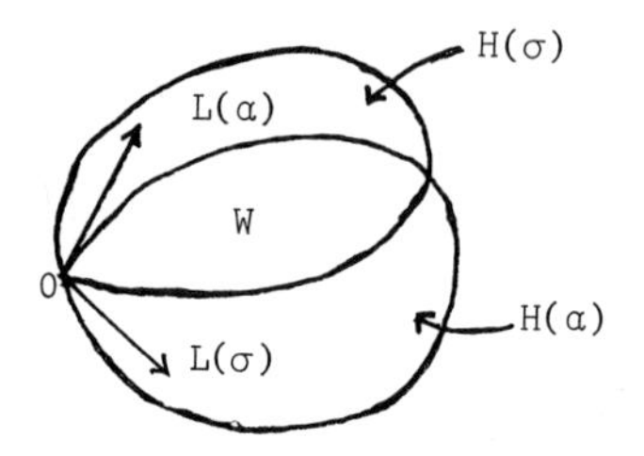

$\dim(W) = n - 2$

$V = L(\alpha) \oplus W \oplus L(\sigma)$.

and it is easy to see $\sigma\alpha = \alpha\sigma$. Therefore

$$\begin{aligned} C &= C_{Ext}(\{\sigma,\beta\}) \\ &= \{\alpha \in Ext \mid L(\alpha) \subset H(\sigma) \cap H(\beta) \text{ and } L \subset H(\alpha)\} \end{aligned}$$

when $L = L(\sigma) = L(\beta)$ for a minimal couple.

Assume $H = H(\sigma) = H(\beta)$. Then, a similar argument gives

$$\begin{aligned} C &= C_{Ext}(\{\sigma,\beta\}) \\ &= \{\alpha \in Ext \mid L(\alpha) \subset H \text{ and } L(\sigma) + L(\beta) \subset H(\alpha)\} . \end{aligned}$$

The above may be summarized as

$$C_{Ext}(\{\sigma,\beta\}) = \{\alpha \in Ext \mid L(\alpha) \subset H(\sigma) \cap H(\beta) \text{ and } L(\sigma) + L(\beta) \subset H(\alpha)\}$$

for a minimal couple $\langle\sigma,\beta\rangle$.

Let $C^2 = C_{Ext}(C_{Ext}(\{\sigma,\beta\}))$ for a minimal couple $<\sigma,\beta>$. We also denote this set as $C^2_{Ext}(\{\sigma,\beta\})$ and call it the <u>double</u> <u>centralizer</u>.

Suppose $L = L(\sigma) = L(\beta)$. Then an argument similar to the above shows that if α is in C^2 then $L(\alpha) = L$ and $H(\sigma) \cap H(\beta) \subset H(\alpha)$. On the other hand, if $H = H(\sigma) = H(\beta)$ then α is in C^2 if $H = H(\alpha)$ and and $L(\alpha) \subset L(\sigma) + L(\beta)$.

To summarize,

<u>(II.14)</u> <u>PROPOSITION</u>. Let $<\sigma,\beta>$ be a minimal couple. Then

(a) the centralizer of $\{\sigma,\beta\}$ by Ext is

$$\{\alpha \in \text{Ext} \mid L(\alpha) \subset H(\sigma) \cap H(\beta) \quad \text{and} \quad L(\sigma) + L(\beta) \subset H(\alpha)\} ,$$

(b) the double centralizer of $\{\sigma,\beta\}$ by Ext is

$$\{\alpha \in \text{Ext} \mid L(\alpha) \subset L(\sigma) + L(\beta) \quad \text{and} \quad H(\sigma) \cap H(\beta) \subset H(\alpha)\} .$$

The statement (II.14) (a) is true for any pair σ and β of extremal involutions with $\sigma\beta \neq \beta\sigma$.

We pause in our remarks concerning involutions and their centralizers to relate minimal couples to transvections. A linear map $\beta : V \to V$ is a <u>projective</u> <u>transvection</u> if either β or $-\beta$ is a transvection. Thus, β is a projective transvection if

(a) there is a hyperplane H of V satisfying $\beta|_H =$ $(\pm)$ (the identity on H).

(b) for all x in V, $\beta(x) - (\pm)x$ is in H where the appropriate sign $(\pm)$ is assumed.

Our immediate purpose relative to transvections is to show that if $\langle\sigma,\beta\rangle$ is a minimal couple then $\sigma\beta$ is a projective transvection of order R.

First some additional observations on minimal couples $\langle\sigma,\beta\rangle$.

(a) Suppose $L = L(\sigma) = L(\beta)$ and $W = H(\sigma) \cap H(\beta)$. Then there are unimodular vectors a and b with a in $H(\sigma)$ and b in $H(\beta)$ satisfying

$$V = L \oplus \underbrace{W \oplus Ra}_{H(\sigma)} = L \oplus \underbrace{W \oplus Rb}_{H(\beta)} .$$

Then $b = sa + te + w$ where $L = Re$ and w is in W. By subtracting elements of W we may assume $b = sa + te$. By checking the images modulo mV, one finds that s and t are units. Thus, if we replace a by $s^{-1}a$ and e by $t^{-1}e$, we may assume $b = a + e$.

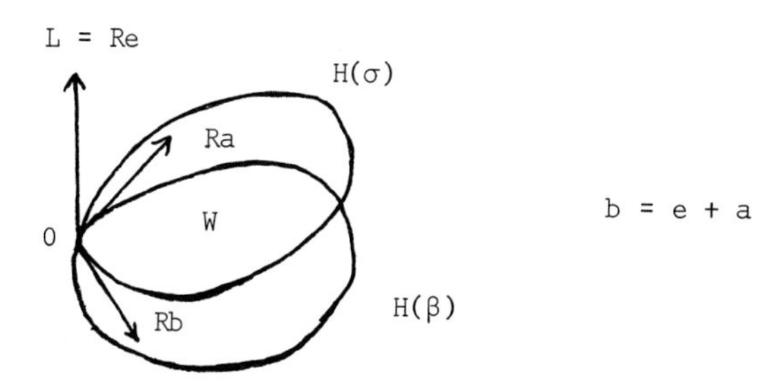

(b) Suppose $H = H(\sigma) = H(\beta)$. Let $L(\sigma) = Ra$ and $L(\beta) = Rb$. Let $H = \oplus \sum Re_i$, $2 \le i \le n$, and $e = e_2$.

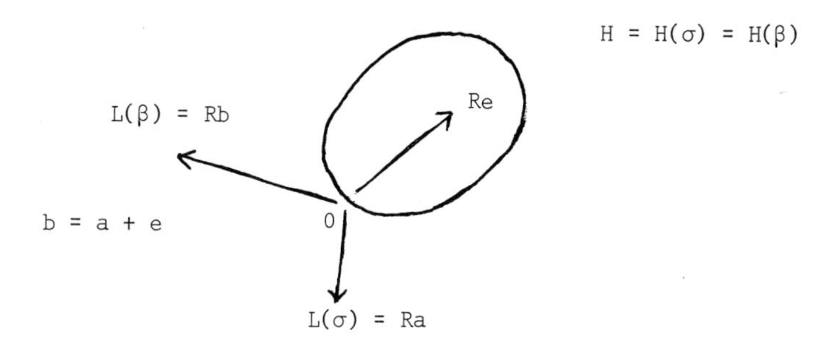

Then, as above, by adjusting e and a appropriately, we may assume $b = a + e$.

(c) Since it is only desired to show $\sigma\beta$ is a projective transvection, both σ and β may be "adjusted" by a multiple of ± 1 . This is a device of Keenan's. Select $\varepsilon = \pm 1$ so that the positive space $P(\varepsilon\sigma)$ of $\varepsilon\sigma$ is the hyperplane $H(\sigma)$ of σ . Similarly, choose $\bar{\varepsilon} = \pm 1$ so that the positive space $P(\bar{\varepsilon}\beta)$ of $\bar{\varepsilon}\beta$ is the hyperplane $H(\beta)$ of β . Then $\varepsilon\sigma\bar{\varepsilon}\beta = \pm\sigma\beta$ and if $\varepsilon\sigma\bar{\varepsilon}\beta$ is a transvection, then $\sigma\beta$ is a projective transvection. Observe $O(\varepsilon\sigma\bar{\varepsilon}\beta) = O(\sigma\beta)$. We denote $\varepsilon\sigma\bar{\varepsilon}\beta$ by τ .

To illustrate the above, suppose $L = L(\sigma) = L(\beta) = Re$. Set $\bar{H} = L \oplus W$ (in (a) above) and $\tau = \varepsilon\sigma\bar{\varepsilon}\beta$. Then $\tau|_{\bar{H}} = id_{\bar{H}}$. Further, to check $\tau(x) - x$ is in $\bar{H}$ we need only check the action of τ on a basis of V . Clearly $\tau(x) - x$ is in $\bar{H}$ for any x in $\bar{H}$. It remains to check the action of τ on a or b . For example,

$$\begin{aligned}
\tau(b) &= (\varepsilon\sigma)(\bar{\varepsilon}\beta)(b) \\
&= (\varepsilon\sigma)(b) \qquad (\text{since } (\bar{\varepsilon}\beta)(b) = b\ ) \\
&= (\varepsilon\sigma)(e + a) \\
&= -e + a \\
&= -e + b - e \\
&= b - 2e \qquad .
\end{aligned}$$

Thus

$$\tau(b) - b = -2e \in Re = L \subset \bar{H} .$$

Since -2 is a unit, $O(\tau) = R$.

Now suppose $H = H(\sigma) = H(\beta)$ and $\tau = \varepsilon\sigma\bar{\varepsilon}\beta$. Obviously, $\tau|_H = id_H$. But $V = H \oplus Rb$. As above, the map τ is a transvection if $\tau(b) - b \in H$.

But

$$\begin{aligned}\tau(b) &= (\varepsilon\sigma)(\bar{\varepsilon}\beta)(b) \\ &= (\varepsilon\sigma)(-b) \qquad (\ L(\beta) = Rb = N(\bar{\varepsilon}\beta)\ ) \\ &= (\varepsilon\sigma)(-a - e) \\ &= (b - e) - e \\ &= b - 2e \quad .\end{aligned}$$

Thus, $\tau(b) - b = -2e$ is in H. Further, -2 is a unit, so $O(\tau) = R$, that is, τ is a unimodular transvection (see Section (II.B)).

(II.15) THEOREM.

(a) Let $\langle\sigma,\beta\rangle$ be a minimal couple. Then $\tau = \sigma\beta$ is a projective unimodular transvection.

(b) Let τ be a non-trivial projective transvection. Then there exist σ and β in Ext with $\tau = \sigma\beta$. Further, if τ is unimodular then we may take $\langle\sigma,\beta\rangle$ to be a minimal couple.

Proof. Part (a) was done in the above paragraphs. To show (b) select a basis $\{e_1,\ldots,e_n\}$ of V such that

$$\mathrm{Mat}(\tau) = \begin{bmatrix} 1 & \alpha & & & \\ 0 & 1 & & & \\ & & 1 & & \\ & & & \ddots & \\ & & & & 1 \end{bmatrix} = \begin{bmatrix} 1 & \alpha & \\ 0 & 1 & \\ & & I \end{bmatrix}$$

(Note: τ is non-trivial.)

(by adjusting τ by ± 1, we may assume τ is a transvection).

Set

$$\mathrm{Mat}(\sigma) = \begin{bmatrix} -1 & 0 & \\ 0 & 1 & \\ & & I \end{bmatrix} , \quad \mathrm{Mat}(\beta) = \begin{bmatrix} -1 & -\alpha & \\ 0 & 1 & \\ & & I \end{bmatrix}$$

relative to $\{e_1,\ldots,e_n\}$. Then $\sigma^2 = 1$, $\beta^2 = 1$ and $\tau = \sigma\beta$. Further

$$L(\sigma) = Re_1$$
$$H(\sigma) = Re_2 \oplus \cdots \oplus Re_n$$

while

$$L(\beta) = Re_1$$
$$H(\beta) = Rf \oplus Re_3 \oplus \cdots \oplus Re_n$$
$$\text{where} \quad f = -\frac{\alpha}{2}e_1 + e_2 .$$

Clearly, σ and β are extremal involutions and $\sigma\beta \neq \beta\sigma$. If τ is unimodular, then $\alpha \not\equiv 0$ modulo m. Thus $\lambda_m\sigma\beta \neq \lambda_m\beta\sigma$. Hence $\langle\sigma,\beta\rangle$ forms a minimal couple.

(II.16) LEMMA. Let $\langle\sigma,\beta\rangle$ be a minimal couple. Let $\bar{\sigma}$ and $\bar{\beta}$ be in the double centralizer of $\{\sigma,\beta\}$, i.e., $C^2_{\mathrm{Ext}}(\{\sigma,\beta\})$. Suppose $\lambda_m\bar{\sigma} \neq \pm\lambda_m\bar{\beta}$. Then

(a) $\langle\bar{\sigma},\bar{\beta}\rangle$ is a minimal couple

(b) $C^2_{Ext}(\{\sigma,\beta\}) = C^2_{Ext}(\{\bar\sigma,\bar\beta\})$.

Proof. (1) Suppose $H = H(\sigma) = H(\beta)$. By (II.14) $H(\bar\sigma) = H(\bar\beta) = H$. Consider $\Pi_m : V \to V/mV$. Clearly $\Pi_m H(\bar\sigma) = \Pi_m H(\bar\beta)$, i.e., $H(\lambda_m\bar\sigma) = H(\lambda_m\bar\beta)$. By $\lambda_m\bar\sigma \neq \pm\lambda_m\bar\beta$, $\Pi_m L(\bar\sigma) \neq \Pi_m L(\bar\beta)$ and consequently $\lambda_m\bar\sigma\bar\beta \neq \lambda_m\bar\beta\bar\sigma$, i.e., $\langle\bar\sigma,\bar\beta\rangle$ is a minimal couple. But, also, if $\Pi_m L(\bar\sigma) \neq \Pi_m L(\bar\beta)$ then $L(\bar\sigma) + L(\bar\beta)$ is a plane. Hence $L(\bar\sigma) + L(\bar\beta) = L(\sigma) + L(\beta)$. Therefore

$$\begin{aligned} C^2_{Ext}(\{\sigma,\beta\}) &= \{\alpha \mid L(\alpha) \subset L(\sigma) + L(\beta) \text{ and } H = H(\alpha)\} \\ &= \{\alpha \mid L(\alpha) \subset L(\bar\sigma) + L(\bar\beta) \text{ and } H = H(\alpha)\} \\ &= C^2_{Ext}(\{\bar\sigma,\bar\beta\}) . \end{aligned}$$

(2) Suppose $L(\sigma) = L(\beta) = L$. Again, by (II.14), $L(\bar\sigma) = L(\bar\beta) = L$. Then $\Pi_m L(\bar\sigma) = \Pi_m L(\bar\beta)$. But $\lambda_m\bar\sigma \neq \pm\lambda_m\bar\beta$. Hence $\Pi_m H(\bar\sigma) \neq \Pi_m H(\bar\beta)$ and $\lambda_m\bar\sigma\bar\beta \neq \lambda_m\bar\beta\bar\sigma$. Thus $\langle\bar\sigma,\bar\beta\rangle$ is a minimal couple. Reducing modulo mV , one also deduces $H(\sigma) \cap H(\beta) = H(\bar\sigma) \cap H(\bar\beta)$. Thus, as above,

$$C^2_{Ext}(\{\sigma,\beta\}) = C^2_{Ext}(\{\bar\sigma,\bar\beta\}) .$$

The goal of this section is the next theorem.

(II.17) THEOREM. Let $\Lambda : GL(V) \to GL(V)$ be a group automorphism and let $\langle\sigma,\beta\rangle$ be a minimal couple. Then $\langle\Lambda\sigma,\Lambda\beta\rangle$ is a minimal couple.

Proof. Let $\langle\sigma,\beta\rangle$ be a minimal couple and suppose $\Sigma = \Lambda\sigma$ and $B = \Lambda\beta$. By (II.12) Σ and B are extremal involutions. By (II.15) there is an $\varepsilon = \pm 1$ such that $\varepsilon\sigma\beta$ is a transvection of order R. Thus, the normal subgroup of $GL(V)$ generated by $\varepsilon\sigma\beta$ is $SL(V)$. (See (II.8).) But $\Lambda\varepsilon = \varepsilon$ and $\Lambda SL(V) = SL(V)$ since $SL(V)$ by (II.6) is the commutator subgroup of $GL(V)$. Further, Λ being a group automorphism preserves group theoretic properties, i.e., the normal subgroup generated by $\varepsilon\sigma\beta$ is carried to the normal subgroup generated by $\varepsilon\Sigma B$. Hence, the normal subgroup generated by $\varepsilon\Sigma B$ is $SL(V)$.

A similar argument, employing the group morphism $\lambda_m : GL(V) \to GL(V/mV)$, indicates the normal subgroup generated by $\lambda_m(\varepsilon\Sigma B)$ is $SL(V/mV)$. Thus, obviously $\lambda_m(\varepsilon\Sigma B) \neq \pm I$, i.e., $\lambda_m\Sigma \neq \lambda_m B$.

If τ_1 and τ_2 are any two extremal involutions with $\tau_1\tau_2 \neq \tau_2\tau_1$ then it is straightforward to show that an extremal involution T commutes with τ_i, $i = 1,2$, if and only if

$$L(\tau_i) \subset H(T)$$
$$L(T) \subset H(\tau_i)$$

for $i = 1,2$ — see remark after (II.14).

Observe (II.14)(a) was stated for minimal couples but could have been formulated for arbitrary non-commuting pairs of extremal involutions.

Since $\sigma\beta \neq \beta\sigma$ we have $\Sigma B \neq B\Sigma$ and by the above remark

$$C_{Ext}(\{\Sigma,B\}) = \{T \in Ext \mid L(T) \subset H(\Sigma) \cap H(B) \text{ and } L(\Sigma) + L(B) \subset H(T)\} .$$

We have $\lambda_m\Sigma \neq \pm\lambda_m B$. Thus, if either $L(\Sigma) = L(B)$ or $H(\Sigma) = H(B)$ then $\langle\Sigma,B\rangle$ is a minimal couple. Hence, if $H(\Sigma) = H(B)$ or $L(\Sigma) = L(B)$, we are done.

Again, $\lambda_m\Sigma \neq \pm\lambda_m B$. Thus, modulo mV , $\Pi_m H(\Sigma) \neq \Pi_m H(B)$ or $\Pi_m L(\Sigma) \neq \Pi_m L(B)$ or both.

Suppose $\Pi_m H(\Sigma) \neq \Pi_m H(B)$. Let $W = H(\Sigma) \cap H(B)$, $L(\Sigma) = Ra$, $L(B) = Rb$ and choose f unimodular in W with $H(B) = W \oplus R(a + \delta f)$.

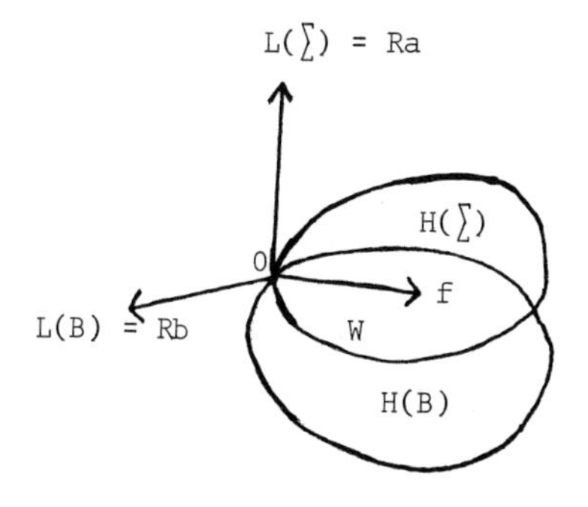

Select an extremal involution $\bar{\Sigma}$ by

$$L(\bar{\Sigma}) = Ra$$
$$H(\bar{\Sigma}) = W \oplus R(a + f)$$

(this determines $\bar{\Sigma}$ up to ± 1). Then

$$L(\bar{\Sigma}) \subset L(\Sigma) + L(B)$$
$$H(\bar{\Sigma}) \supset H(\Sigma) \cap H(B)$$

and, consequently, $\bar{\Sigma}$ commutes with each element in $C_{Ext}(\{\Sigma,B\})$, i.e., $\bar{\Sigma}$ is in $C^2_{Ext}(\{\Sigma,B\})$.

Clearly, $<\Sigma,\bar{\Sigma}>$ is a minimal couple.

Suppose the line $L(B) \neq L(\Sigma)$. Then B is not in $C^2_{Ext}(\{\Sigma,\bar{\Sigma}\})$ by (II.14). Hence $C^2_{Ext}(\{\Sigma,\bar{\Sigma}\}) \neq C^2_{Ext}(\{\Sigma,B\})$.

The centralizer of a subset X of $GL(V)$ under a group automorphism $\Phi : GL(V) \to GL(V)$ is carried to the centralizer of ΦX since being a centralizer is a group theoretic property. Taking $\Phi = \Lambda^{-1}$, we have

$$C^2_{Ext}(\{\sigma,\bar{\sigma}\}) \neq C^2_{Ext}(\{\sigma,\beta\})$$

where $\bar{\sigma} = \Lambda^{-1}(\bar{\Sigma})$. By an argument similar to the initial part of this proof, since $<\Sigma,\bar{\Sigma}>$ is a minimal couple, one concludes

$$\lambda_m \sigma \neq \pm \lambda_m \bar{\sigma} .$$

Thus, by (II.16),

$$C^2_{\mathrm{Ext}}(\{\sigma,\bar{\sigma}\}) = C^2_{\mathrm{Ext}}(\{\sigma,\beta\}) .$$

But this is a contradiction to the assumption $L(B) \neq L(\Sigma)$. Hence $L(B) = L(\Sigma)$. This completes the first case.

Suppose $\Pi_m L(\Sigma) \neq \Pi_m L(B)$. Further, assume $\Pi_m H(\Sigma) = \Pi_m H(B)$ (for otherwise, we are in the above case).

Set
$$L(\Sigma) = Re$$
$$H(\Sigma) = Ra_2 \oplus \cdots \oplus Ra_n .$$

Since $\Pi_m H(\Sigma) = \Pi_m H(B)$, there exist $\alpha_2,\ldots,\alpha_n$ in m with

$$H(B) = R(a_2 + \alpha_2 e) \oplus \cdots \oplus R(a_n + \alpha_n e) .$$

Let $L(B) = R(\beta_1 e + \beta_2 a_2 + \cdots + \beta_n a_n)$. Since $\Pi_m L(B) \not\subset \Pi_m H(B)$, β_1 is a unit.

Set
$$a = \beta_1 e$$
$$f = \beta_2 a_2 + \cdots + \beta_n a_n .$$

Then $\Pi_m f \neq 0$ since $\Pi_m L(B) \neq \Pi_m L(\Sigma)$ we have

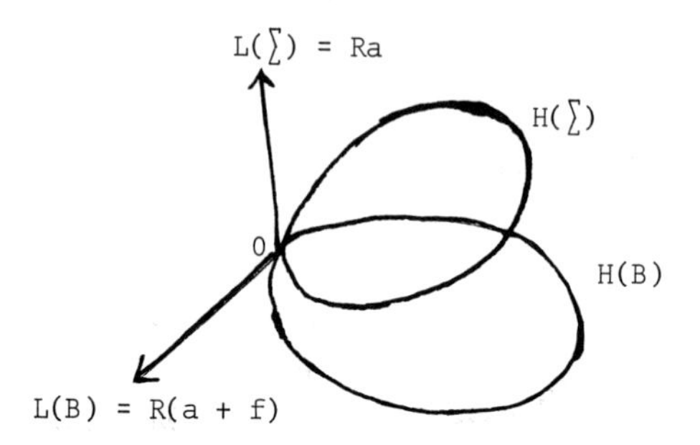

Define an extremal involution $\bar{\Sigma}$ by

$$L(\bar{\Sigma}) = L(B) = R(a + f)$$
$$H(\bar{\Sigma}) = H(\Sigma) \quad .$$

Then $\langle\Sigma,\bar{\Sigma}\rangle$ is a minimal couple.

<u>Suppose</u> $H(B) \neq H(\Sigma)$.

Then as in the above proof, Σ and $\bar{\Sigma}$ are in $C^2_{Ext}(\{\Sigma,B\})$. But B is not in $C^2_{Ext}(\{\Sigma,\bar{\Sigma}\})$. We arrive at a contradiction to the assumption $H(B) \neq H(\Sigma)$. Hence $H(B) = H(\Sigma)$.

Thus $\langle\Sigma,B\rangle = \langle\Lambda\sigma\ \Lambda\beta\rangle$ is a minimal couple.

(F) THE AUTOMORPHISMS OF GL(V)

We continue to assume that V is a space over a local ring R of dimension ≥ 3. Further, 2 is a unit of R, m denotes the maximal ideal of R and $k = R/m$.

In (I.D) we introduced the concept of a σ-semi-linear morphism $\phi : V \to V$. The range space V may have been taken to be any R-space. Thus $\phi : V \to W$ is a σ-semi-linear morphism if

(a) $\phi(v_1 + v_2) = \phi(v_1) + \phi(v_2)$

(b) $\phi(rv) = \sigma(r)\phi(v)$

for v, v_1, v_2 in V, r in R and $\sigma : R \to R$ a ring isomorphism. In this section W will be taken as either V or V^*.

If $g : V \to V$ is a semi-linear isomorphism of V onto V, we define $\Phi_g : GL(V) \to GL(V)$ by

$$\Phi_g(\beta) = g\beta g^{-1}$$

for all β in $GL(V)$. The map Φ_g is an automorphism of $GL(V)$. The maps Φ_g form our first class of automorphisms of $GL(V)$.

Let $\bar{B}$ be a basis of V. Suppose the matrix of g is G and the associated ring automorphism of g is σ. Then, if B is the matrix of β relative to $\bar{B}$,

$$\Phi_g : B \to GB^{\sigma}G^{-1}$$

where $B^{\sigma} = [\sigma b_{ij}]$ if $B = [b_{ij}]$.

(II.18) LEMMA.

(a) $\Phi_g\Phi_h = \Phi_{gh}$.

(b) $\Phi_g^{-1} = \Phi_{g^{-1}}$.

(c) Φ_g induces a group automorphism on $GC(V,m^t)$ and $SC(V,m^t)$ for $t = 0,1,\ldots$.

(d) For β in $GL(V)$,

$$\det(\Phi_g(\beta)) = \sigma(\det(\beta)) .$$

(e) $\Phi_g = \Phi_h$ if and only if $g = \alpha h$, α unit in R .

Proof of (c). If β is in $GC(V,m^t)$ then $\lambda_{m^t}(\beta)$ is in the center of $GL(V/m^tV)$. Since ring automorphisms $\sigma : R \to R$ carry m^t into m^t , a direct computation shows $\lambda_{m^t}(\Phi_g(\beta))$ is in the center of $GL(V/m^tV)$. Hence $\Phi_g(\beta)$ is in $GC(V,m^t)$. To show the second part of the statement, recall (II.6), $SC(V,m^t) = [GL(V),GC(V,m^t)]$.

Recall from (I.F), we have a natural pairing

$$(\ , \ ) : V^* \times V \to R$$

by

$$(f,v) = f(v)$$

for f in V^* and v in V . If $\sigma : V \to V$ is an R-morphism, then the <u>transpose</u> or <u>adjoint</u> of σ , denoted σ^* , is defined by

$$(f\sigma^*, v) = (f, \sigma(v)) .^1$$

Thus $\sigma^* : V^* \to V^*$ and for σ and β in End(V) , $(\sigma\beta)^* = \beta^*\sigma^*$. Suppose a basis B of V is fixed and B^* denotes the dual basis of V^* obtained from B (see (I.C)). Then, if $\mathrm{Mat}(\sigma) = [a_{ij}]$ relative to B , we have $\mathrm{Mat}(\sigma^*) = [b_{ij}]$ where $b_{ij} = a_{ji}$ for $1 \le i,j \le n$. That is, $\mathrm{Mat}(\sigma^*) = \mathrm{Mat}(\sigma)^t$ is the transpose of matrices.

If σ is in GL(V) then the <u>contragredient</u> $\check{\sigma}$ of σ is defined by

$$\check{\sigma} = (\sigma^*)^{-1} .$$

Note $(\sigma^*)^{-1} = (\sigma^{-1})^*$ and $\widecheck{\beta\sigma} = \check{\beta}\check{\sigma}$.

With the aid of the contragredient, we define a second class of automorphisms of GL(V) .

Let $h : V \to V^*$ be a μ-semi-linear isomorphism of V onto V^* . Define $\Psi_h : GL(V) \to GL(V)$ by

$$\Psi_h(\beta) = h^{-1}\check{\beta}h$$

for all β in GL(V) . It is easy to verify that Ψ_h is an automorphism of GL(V) .

Let H be the matrix of h relative to a basis $\bar{B}$ of V with dual

---

[1] The transpose will be written acting on the left or right depending on convenience, e.g., see (III.A).

$\bar{B}^*$ of V^*. Suppose B is the matrix of β relative to $\bar{B}$. Then the matrix of $\Psi_h(\beta)$ is

$$(H^{-1}\check{B}H)^{\mu^{-1}}$$

relative to $\bar{B}$. Here $\check{B} = (B^t)^{-1}$.

(II.19) LEMMA.

(a) $\Phi_g \Psi_h = \Psi_{hg^{-1}}$, $\Psi_h^{-1} = \Psi_{h^{-1}}$.

(b) Ψ_h induces a group automorphism on $GC(V,m^t)$ and $SC(V,m^t)$ for $t = 0,1,2,\ldots$.

(c) For β in $GL(V)$,

$$\det(\Psi_h(\beta)) = \mu^{-1}(\det(\beta^{-1})) .$$

(d) $\Psi_g = \Psi_h$ if and only if $g = h\alpha$ for α a unit in R.

A third class of automorphisms of $GL(V)$ are the radial automorphisms. An automorphism $\Lambda : GL(V) \to GL(V)$ is called a radial automorphism of $GL(V)$ if there is a group morphism

$$\chi : GL(V) \to RL(V) = \text{center of } GL(V)$$

satisfying

$$\Lambda(\beta) = \chi(\beta)\beta$$

for all β in $GL(V)$. A radial automorphism Λ determines a unique χ, thus Λ is denoted by

$$P_\chi \ .$$

Observe an arbitrary group morphism

$$\eta : GL(V) \to RL(V) \ ,$$

for example, $\beta \to \det(\beta)$, does not, in general, determine a radial automorphism.

However, the determinant does appear in χ . Since χ is a group morphism of $GL(V)$ into an Abelian group $RL(V)$, χ factors through $GL(V)$ modulo its commutator subgroup $[GL(V),GL(V)]$. But, by (II.6), $SL(V) = [GL(V),GL(V)]$. We have

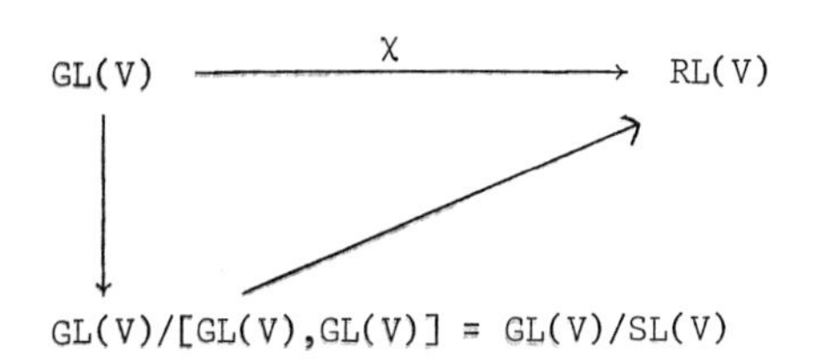

.

But $GL(V)/SL(V) \simeq R^*$ under $\beta SL(V) \to \det(\beta)$ (see (II.7) or (I.10) and (I.11)). Consequently, χ may be viewed as a composition of $\det : GL(V) \to R^*$ and a group morphism $\theta : R^* \to RL(V)$, i.e.,

$$\chi(\beta) = \theta(\det(\beta)) \ .$$

The purpose of this section is to verify our earlier claim on the form of Λ . That is, if $\Lambda : GL(V) \to GL(V)$ is a group automorphism then Λ may be expressed as either

$$\Lambda = P_\chi \circ \Phi_g \qquad \text{(Automorphism of the \underline{First Type})}$$

or

$$\Lambda = P_\chi \circ \Phi_h \qquad \text{(Automorphism of the \underline{Second Type})} .$$

Suppose $\mathcal{T}$ is a family of extremal involutions. The elements of $\mathcal{T}$ are said to <u>share a line</u> L if $L = L(\sigma)$ for all σ in $\mathcal{T}$. Similarly, the elements of $\mathcal{T}$ <u>share a hyperplane</u> H if $H = H(\sigma)$ for all σ in $\mathcal{T}$. If all the elements of $\mathcal{T}$ share either a hyperplane or a line, we say they <u>share a space</u>.

Let W be either a hyperplane or a line in V. Set

$$\mathcal{T}(W) = \{\sigma \text{ in Ext} \mid H(\sigma) = W \text{ or } L(\sigma) = W\}$$

Thus, $\mathcal{T}(W)$ is the collection of all extremal involutions which share the space W.

Select σ in $\mathcal{T}(W)$. It is easy to construct τ and β in $\mathcal{T}(W)$ with $\langle\sigma,\tau\rangle$, $\langle\sigma,\beta\rangle$, $\langle\tau,\beta\rangle$ minimal couples. If σ,τ,β in $\mathcal{T}(W)$ satisfy $\langle\sigma,\tau\rangle$, $\langle\sigma,\beta\rangle$ and $\langle\tau,\beta\rangle$ being minimal couples, then we say σ,τ,β are <u>simplicially linked</u>. Pictorially,

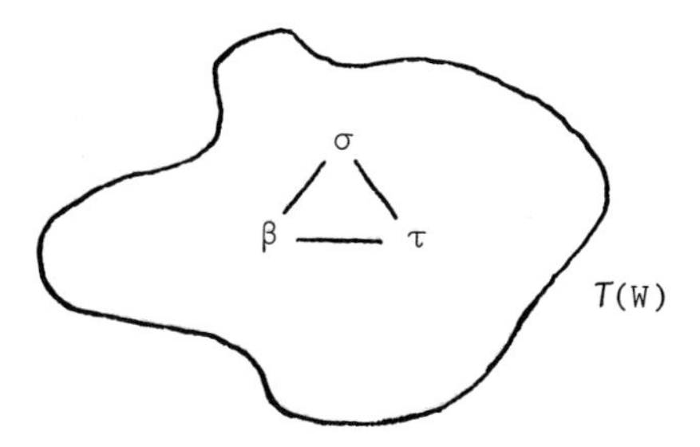

Two observations:

(1) If $\Lambda : GL(V) \to GL(V)$ is a group automorphism and σ,β,τ are simplicially linked in $T(W)$, then, applying (II.17), $\Lambda\sigma,\Lambda\beta,\Lambda\tau$ are simplicially linked. That is, $\Lambda\sigma$, $\Lambda\beta$ and $\Lambda\tau$ share the same space.

(2) If μ is any other element of $T(W)$, then μ is simplicially linked to at least one of the sides of the above triangle.

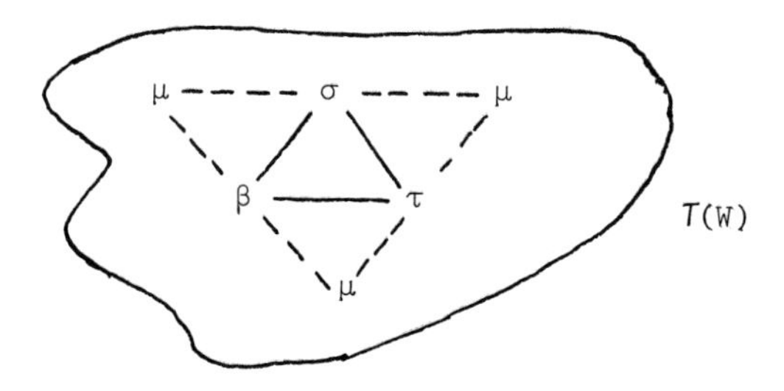

That is, $<\mu,\alpha_1>$ and $<\mu,\alpha_2>$ are minimal couples for some choice of α_1,α_2 in $\{\sigma,\beta,\tau\}$.

Let
$$T(W)^\wedge = \{\wedge\sigma \mid \sigma \text{ is in } T(W)\} .$$

<u>(II.20)</u> <u>LEMMA</u>. The elements of $T(W)^\wedge$ share the same space.

<u>Proof</u>. This follows from the observations above.

Denote the space shared by the elements of $T(W)^\wedge$ as
$$W^\wedge .$$

<u>(II.21)</u> <u>LEMMA</u>. Let W_1 and W_2 be either a pair of hyperplanes or a pair of lines in V. Then
$$\dim(W_1^\wedge) = \dim(W_2^\wedge) .$$

<u>Proof</u>. Select ρ in $GL(V)$ with $\rho W_1 = W_2$. Then, by (II.10),
$$T(W_2) = \rho T(W_1)\rho^{-1} .$$

Thus,
$$T(W_2)^\wedge = \wedge\rho(T(W_1)^\wedge)\wedge\rho^{-1}$$

and

$$\dim(W_2^\wedge) = \dim(W_1^\wedge) .$$

Suppose W, in the above discussion, is a line L and $\wedge$ is an automorphism of $GL(V)$. Then, $\wedge$ induces a natural map $P(V) \to \bar{P}(V)$ by $L \to L^\wedge$ where $\bar{P}(V)$ equals either $P(V)$ or the collection of hyperplanes of V.

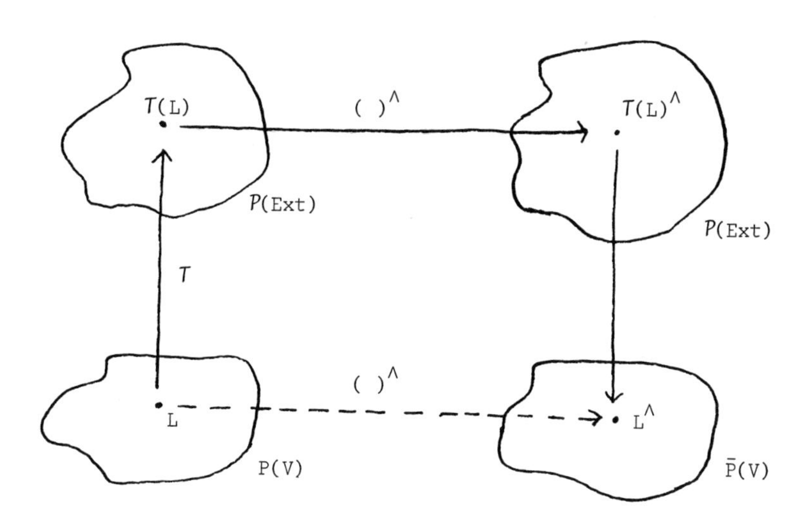

where $\mathcal{P}(Ext)$ denotes the set of subsets of Ext .

Case I. $L^\wedge$ is a line.

Suppose $L^{\wedge}$ is a line. Then the map

$$(\ )^{\wedge} : L \to L^{\wedge}$$

induced by the automorphism $\wedge$ carries lines to lines, i.e., carries $P(V)$ to $P(V)$. It is easy to see $(\ )^{\wedge}$ is bijective.

Suppose L , L_1 and L_2 are lines in V with $L \subset L_1 \oplus L_2$ (see definition of projectivity prior to (I.13)). Let $V = L_1 \oplus L_2 \oplus W$. Let σ and β be extremal involutions with positive spaces

$$P(\sigma) = L_2 \oplus W$$
$$P(\beta) = L_1 \oplus W$$

and negative spaces

$$N(\sigma) = L_1$$
$$N(\beta) = L_2 .$$

Set $\alpha = \sigma\beta$. Then

$$N(\alpha) = L_1 \oplus L_2$$
$$P(\alpha) = W .$$

Consider the following subset of the centralizer of $\{\sigma,\beta\}$ in Ext ,

$$C^*_{Ext}(\sigma,\beta) = \{\delta \text{ in } Ext \mid \delta \text{ in } C_{Ext}(\sigma,\beta) \text{ but } \delta \neq \pm\sigma \ , \ \delta \neq \pm\beta\} .$$

Then

$$C^*_{Ext}(\sigma,\beta) = \{\delta \text{ in Ext} \mid L(\delta) \subset W, L_1 \oplus L_2 \subset H(\delta)\} .$$

Since $L \subset L_1 \oplus L_2$, we have $L_1 \oplus L_2 = L \oplus K$ where K is a line. Select δ , an extremal by

$$L(\delta) = L$$
$$H(\delta) = K \oplus W .$$

Then δ commutes with α but δ is <u>not</u> in $C^*_{Ext}(\sigma,\beta)$.

Apply the automorphism $\wedge : GL(V) \to GL(V)$ to the above.

Let

$$A = \wedge\alpha = \wedge\sigma\beta = \wedge\sigma\wedge\beta$$
$$\textstyle\sum = \wedge\sigma , \quad L(\wedge\sigma) = L(\sum) = L_1^\wedge$$
$$B = \wedge\beta , \quad L(\wedge\beta) = L(B) = L_2^\wedge$$
$$\Delta = \wedge\delta , \quad L(\wedge\delta) = L(\Delta) = L^\wedge .$$

Since $\sum$ and B are commuting, distinct involutions with $\sum \neq \pm B$, we have $V = L_1^\wedge \oplus L_2^\wedge \oplus \bar{W}$. Hence $L_1^\wedge + L_2^\wedge$ is a plane and the sum is direct.

The involution A has spaces $L_1^\wedge \oplus L_2^\wedge$ and $H(\sum) \cap H(B)$. Since Δ commutes with A either

$$L(\Delta) \subset L_1^\wedge + L_2^\wedge \quad \text{and} \quad H(\Delta) \supset H(\Sigma) \cap H(B)$$

or

$$L(\Delta) \subset H(\Sigma) \cap H(B) \quad \text{and} \quad L_1^\wedge + L_2^\wedge \subset H(\Delta) .$$

However, if the latter situation occurs, then Δ commutes with Σ and B which implies δ commutes with σ and β — a contradiction.

Thus, $L(\Delta) \subset L_1^\wedge + L_2^\wedge$, i.e.,

$$L^\wedge \subset L_1^\wedge \oplus L_2^\wedge$$

and the map

$$L \to L^\wedge$$

is a projectivity $P(V) \to P(V)$.

<u>Case II</u>. $L^\wedge$ is a hyperplane.

We use the hyperplane $L^\wedge$ to determine a line in the dual space V^* of V . This is done as follows. Define a pairing (as in (I.F) or (III.A))

$$(\ ,\ ) : V^* \times V \to R$$

by

$$(f,v) = f(v) .$$

If W is any subspace of V , define

$$\begin{aligned} W^\perp &= \{f \text{ in } V^* \mid (f,W) = 0\} \\ &= \{f \text{ in } V^* \mid (f,w) = 0 \text{ for all } w \text{ in } W\} . \end{aligned}$$

Thus $W^{\perp}$ consists of elements f in W^* which annihilate W and is called the <u>orthogonal complement of</u> W <u>in</u> V^*. It is straightforward to verify (e.g., look ahead to (III.17) — recall also (I.F))

(a) $W^{\perp}$ is a subspace of V^*,

(b) $\dim(W^{\perp}) = \dim(V/W)$,

(c) the map $W \to W^{\perp}$ inverts the partial order on subspaces of V.

We apply the above to the automorphism $\wedge : GL(V) \to GL(V)$. Thus

$$L \to L^{\wedge} \to (L^{\wedge})^{\perp}$$

gives

$$P(V) \to \{\text{hyperplanes of } V\} \to P(V^*) .$$

It is easy to check that the composition is bijective; if L, L_1 and L_2 are lines in $P(V)$ with $L \subset L_1 \oplus L_2$ then $L_1^{\wedge} \cap L_2^{\wedge} \subset L^{\wedge}$ and $\Pi L_1^{\wedge} \neq \Pi L_2^{\wedge}$; if H, H_1, H_2 are hyperplanes with $\Pi H_1 \neq \Pi H_2$ and $H_1 \cap H_2 \subset H$ then $H^{\perp} \subset H_1^{\perp} \oplus H_2^{\perp}$.

Let $(L^{\wedge})^{\perp}$ be denoted by ${}^{\wedge}L$. We conclude the map

$$L \to {}^{\wedge}L$$

is a projectivity

$$P(V) \to P(V^*) .$$

The Fundamental Theorem of Projective Geometry as stated and proven in (I.13) represented projectivities $\alpha : P(V) \to P(V)$ as being induced from semi-linear isomorphisms $\phi : V \to V$. However, with a wave of the hand, we claim a quick glance at the proof shows the image may be taken to be $P(W)$ where W is any space of dimension the same as V. Indeed, more generally, one could show the following, employing the same proof (with minor modifications) of (I.13):

<u>(II.22)</u> <u>THEOREM</u>. (Fundamental Theorem of Projective Geometry) Let R_1 and R_2 be local rings. Let V_i be an R_i-space for $i = 1,2$. Suppose

$$\alpha : P(V_1) \to P(V_2)$$

is a projectivity. Then, if $\dim(V_1) \geq 3$ the following are true:

(a) $\dim(V_1) = \dim(V_2)$.

(b) There is a ring isomorphism $\sigma : R_1 \to R_2$.

(c) There is a σ-semi-linear isomorphism $\phi : V_1 \to V_2$ with $P(\phi) = \alpha$.

Actually, our principal interest in the above result lies only in application to projectivities $P(V) \to P(V)$ and $P(V) \to P(V^*)$.

<u>(II.23)</u> <u>LEMMA</u>. Let $\Gamma : GL(V) \to GL(V)$ be a group automorphism. Suppose for each line L in V and σ in $\mathcal{T}(L) = \{\beta \text{ in Ext} \mid L(\beta) = L\}$ we have $L(\sigma)^\Gamma = L(\sigma)$. Then, there is a group morphism χ of $GL(V)$ into its center satisfying

$$\Gamma(\alpha) = \chi(\alpha)\alpha$$

for all α in $GL(V)$. Thus $\Gamma = P_\chi$.

Proof. Let α be in $GL(V)$, L a line in V and σ in $T(L)$. Then $(\Gamma\alpha)(\Gamma\sigma)(\Gamma\alpha)^{-1}$ is an extremal involution with line $(\Gamma\alpha)L$ (see (II.10)). On the other hand, $\Gamma(\alpha\sigma\alpha^{-1})$ is an extremal involution with line αL . Hence $(\Gamma\alpha)L = \alpha L$ and $(\alpha^{-1}\Gamma\alpha)L = L$. Let $L = Re$. Then $(\alpha^{-1}\Gamma\alpha)(e) = r_e e$ since $\alpha^{-1}\Gamma\alpha$ is R-linear and $\alpha^{-1}\Gamma\alpha$ maps L to L . Similarly, if f is unimodular, $\alpha^{-1}\Gamma\alpha(f) = r_f f$.

Suppose e and f are R-free. Then $e + f$ is unimodular and

$$r_{e+f}(e + f) = r_e e + r_f f .$$

Thus, $r_e = r_{e+f} = r_f$. Letting $\chi(\alpha)$ be r_e , we have $\alpha^{-1}\Gamma\alpha(e) = \chi(\alpha)e$ for all e unimodular in V . Further $\chi : \alpha \to \chi(\alpha)$ is a map $GL(V) \to \text{Center}(GL(V))$. It is easy to check χ is a group morphism.

To apply (II.23) in the general situation we treat, as before, two cases.

Suppose $\Lambda : GL(V) \to GL(V)$ is a group automorphism. Let L be any line in V and $T(L) = \{\sigma \text{ in Ext} \mid L(\sigma) = L\}$.

Case I. Suppose $L^\Lambda = L$.

By the earlier discussion of Case I, the map $P(V) \to P(V)$ by $L \to L^{\wedge}$ is a projectivity. By (I.13) let $g : V \to V$ be a semi-linear isomorphism with $P(g)L = L^{\wedge}$, i.e., $gL = L^{\wedge}$. Construct $\Phi_g : GL(V) \to GL(V)$ by $\Phi_g(\beta) = g\beta g^{-1}$. If σ is an extremal involution in $T(L)$, then $\Phi_g(\sigma)$ is an involution with line gL (see (II.10)). Thus, if $\Gamma = \wedge\Phi_g^{-1}$ then

$$L^{\Gamma} = L^{\wedge\Phi_g^{-1}} = (gL)^{\Phi_g^{-1}} = g^{-1}(gL) = L .$$

Applying (II.23),

$$\Gamma(\alpha) = \chi(\alpha)\alpha \qquad \alpha \text{ in } GL(V) .$$

That is $\wedge\Phi_g^{-1} = P_\chi$ or

$$\wedge = P_\chi \circ \Phi_g .$$

<u>Case</u> <u>II</u>. Suppose $L^{\wedge}$ is a hyperplane.

The argument is similar to the one used above. By the earlier discussion of Case II, the map $L \to {}^{\wedge}L$ of $P(V)$ into $P(V^*)$ is a projectivity. Thus, by (II.22) we obtain a semi-linear isomorphism $h : V \to V^*$ which induces this projectivity and a group automorphism $\Psi_h : GL(V) \to GL(V)$.

If H denotes the hyperplane $L^{\wedge}$ for a line L in V, then ${}^{\wedge}L = H^{\perp}$ and if σ is in $T(L)$ then the line $L(\check{\sigma})$ is $H^{\perp}$. Further, the line of $\Psi_h\sigma$ is $h^{-1}(H^{\perp})$. This gives an automorphism $\Psi_h\wedge : GL(V) \to GL(V)$.

Lemma (II.23) applies to this automorphism and we obtain

$$\Lambda^{-1}(\alpha) = \chi(\alpha)\Psi_h(\alpha)$$

and, consequently,

$$\Lambda = P_\chi \circ \Psi_{\underline{h}}$$

where $\underline{h} = \check{h}^{-1}$. We have proven:

<u>(II.24)</u> <u>THEOREM</u>. (Characterization of Automorphisms of GL(V))
Let R be a local ring having 2 a unit and let V be an R-space with $\dim(V) \geq 3$. If

$$\Lambda : GL(V) \to GL(V)$$

is a group automorphism, then either

$$\Lambda = P_\chi \circ \Phi_g \qquad \text{(Automorphism of First Type)}$$

or

$$\Lambda = P_\chi \circ \Psi_h \qquad \text{(Automorphism of Second Type)}$$

for suitable P_χ and Φ_g or Ψ_h (as described in this section).

There remain questions of uniqueness of representation and some corollaries.

<u>(II.25)</u> <u>PROPOSITION</u>. (Uniqueness for Automorphisms of First Type)
Suppose P_{χ_1}, P_{χ_2}, Φ_{g_1} and Φ_{g_2} are given automorphisms of GL(V) .

Then the following are equivalent.

(a) $P_{\chi_1} \circ \Phi_{g_1} = P_{\chi_2} \circ \Phi_{g_2}$.

(b) $P_{\chi_1} = P_{\chi_2}$ and $\Phi_{g_1} = \Phi_{g_2}$.

(c) $\chi_1 = \chi_2$ and $g_1 = \alpha g_2$ for α in $RL(V)$.

Proof. We show (a) implies (c). The implications, (c) implies (b) and (b) implies (a), are immediate. Let L be a line in V . Let $g = g_1 g_2^{-1}$. Then $P_{\chi_2} = P_{\chi_1} \circ \Phi_g$. But $L^{P_{\chi_2}} = L$. Thus $L^{P_{\chi_1} \circ \Phi_g} = gL = L$. Thus, for each line L in V , $gL = L$ and consequently $g = \alpha$ in $RL(V)$ and $g_1 = \alpha g_2$. Then $\Phi_{g_1} = \Phi_{g_2}$ which implies $P_{\chi_1} = P_{\chi_2}$, hence $\chi_1 = \chi_2$.

(II.26) PROPOSITION. (Uniqueness for Automorphisms of Second Type) Suppose P_{χ_1} , P_{χ_2} , Ψ_{h_1} and Ψ_{h_2} are given automorphisms. Then the following are equivalent.

(a) $P_{\chi_1} \circ \Psi_{h_1} = P_{\chi_2} \circ \Psi_{h_2}$.

(b) $P_{\chi_1} = P_{\chi_2}$ and $\Psi_{h_1} = \Psi_{h_2}$.

(c) $\chi_1 = \chi_2$ and $h_1 = h_2 \alpha$ for α in $RL(V)$.

Proof. The proof is similar to (II.25).

Finally, it is easy to verify that

$$P_{\chi_1} \circ \Phi_g \neq P_{\chi_2} \circ \Psi_h$$

for P_{χ_1} , P_{χ_2} , Φ_g and Ψ_h by checking the projectivities induced by these automorphisms.

A subgroup H of a group G is characteristic if $\Lambda(H) \subset H$ for every automorphism Λ of G .

(II.27) PROPOSITION. Let R be a local ring with maximal ideal m and having 2 a unit. If $\dim(V) \geq 3$ then $GC(V,m^t)$ and $SC(V,m^t)$ are characteristic subgroups of $GL(V)$.

Proof. If $\mu : R \to R$ is a ring automorphism then $\mu(m^t) \subseteq m^t$. Thus if σ is in $GC(V,m^t)$ then $\Phi_g(\sigma) = g\sigma g^{-1}$ is in $GC(V,m^t)$. Consequently, $(P_\chi \circ \Phi_g)(\sigma)$ is in $GC(V,m^t)$. Since, by (II.6), $SC(V,m^t) = [GL(V),GC(V,m^t)]$, $SC(V,m^t)$ is characteristic under $P_\chi \circ \Phi_g$ and the congruence subgroups are characteristic under automorphisms of the first type. The case for automorphisms of the second type is handled similarly.

(G) DISCUSSION

The origin of the results in the chapter may be traced to the classification problems of simple groups, namely, "What are the simple groups?" equivalently "Determine the normal subgroups of a group." and "When are two simple groups isomorphic?" The problem for linear groups was first to determine maximal normal subgroups of the classical linear groups and thus, from their quotient groups, create simple groups. Then, second, determine the isomorphisms between these linear groups, to discover which of the simple groups created in the above fashion were the same. In 1928, Schreier and Van der Waerden [132] initiated the isomorphism theory of classical groups by determining the isomorphisms of the projective special linear group over a field.

Probably the first detailed examination of the normal subgroups of GL(V) over a ring R which was not a field was undertaken by Brenner ([47], [48] and [49]) where R was the ring of rational integers or its residue class rings.

The boundaries of the subject, e.g., "What is geometric algebra?" were established somewhat later by Artin, Geometric Algebra, (1957), and Dieudonné, On the Automorphisms of Classical Groups, (1951), in the context of vector spaces over division rings.

We first discuss the normal subgroups of GL(V) .

After Brenner's early work, in 1961, Klingenberg [89] characterized the normal subgroups of GL(V) when R was local, 2 a unit and $\dim(V) \geq 3$. It should be clear that low dimensions create added difficulties. Roughly, from a matrix theoretic approach, low dimensions do not provide sufficient off-diagonal positions or, geometrically, there is not a sufficient number of distinct lines and planes. A similar situation occurs in convexity theory where problems related to convex shapes in the plane vanish in 3-space. Lacroix in 1969 [96] examined the normal subgroups of GL(V) over a local ring when dim(V) = 2 .[1]

The above would be considered a "ring theoretic" approach to normal subgroup theory, i.e., the field is replaced by a suitable commutative or non-commutative ring. Simultaneously, there is a "number theoretic" variation where the field is first replaced by the rational integers and then by a ring of algebraic integers, arithmetic subrings of fields and, if possible, orders.

The prime mover of the general theory has been Hyman Bass. Building on the idea that increased dimension permits more "movement" of planes, lines and group elements, Bass sought to show stable behavior occurred for normal subgroups in higher dimension. In 1964, [37] Bass defined the "stable range" of a ring R and proved that if dim(V) exceeds the stable range of R , then the normal subgroups are caught between congruence subgroups, commutator relations behave properly and the

---

[1] See the Appendix of (IV) for a description of the normal subgroups of GL(V) , dim(V) = 2 , when 2 and 3 are units.

group $E(V)$ of elementary transvections is transitive on the collection of unimodular vectors in V. As might be expected from the local ring setting, stability occurs when the dimension of the space is large compared with the dimension of the maximal spectrum of the ring.

From the standpoint of classifying normal subgroups several directions of work are prominent.

(a) A sharp determination of the stable range of a ring. For results in this direction see [66], [70] and [17].

(b) An examination of the case $\dim(V) = 2$. Excluding the above paper by Lacroix, we know of little work in this direction for commutative rings. For rings of integers see [21].

(c) Let $n \to \infty$ and attempt to determine if stable behavior occurs. Roughly, "stable behavior" is characterized by the type of result occurring in this chapter. This is answered affirmatively by Bass [4], [37] for $\varinjlim GL_n(R)$ and by Maxwell [106] for $GL(V)$ when $\dim(V)$ is not finite (V an R-space).

(d) Replace V by a more general module P (probably projective) and determine the normal subgroups of $GL(P)$. The most recent work in this direction is by Bak [35].

A number of fundamental problems concerning $GL(V)$ and $SL(V)$ for an arbitrary ring R are interwoven with the study of normal subgroups. We only briefly sketch a few and do not illustrate their interdependence.

First, there is the "congruence subgroup problem" for $SL(V)$ over a ring R, i.e., "When do the subgroups of $SL(V)$ of finite index contain congruence subgroups?" This has been solved when R is a ring of algebraic integers by Bass, Milnor and Serre [37] and [42]. Mennicke [107] had previously shown this when R was the rational integers and $n \geq 3$ ($n = 2$ is false).

Second, there is the "finite generation problem" for $GL(V)$, i.e., "Under what conditions on R or $\dim(V)$, is $GL(V)$ a finitely generated group?" For a detailed discussion of this problem see Bass [5].

Third, there is from K-theory the "calculation of $K_1(R)$." See [3], [4] or [27]. If

$$GL(R) = \varinjlim GL_n(R) ,$$

$$E(R) = \varinjlim E_n(R)$$

and, when R is commutative,

$$SL(V) = \varinjlim SL_n(R)$$

then

$$K_1(R) = GL(R)/E(R) .$$

If R is commutative, then $\det : GL(R) \to R^*$ splits and

$$K_1(R) = R^* \oplus SK_1(R)$$

where $SK_1(R) = SL(R)/E(R)$. When is $SK_1(R)$ trivial? Bass [5] has shown there exist principal ideal domains for which $SK_1(R)$ is non-trivial. This is related to the question mentioned in (I.G) of how the group of elementary matrices relative to a fixed basis sit in the special linear group.

Fourth, "When are projective modules free?" If E denotes the set of unimodular elements of V ($\dim(V) = n$) then $GL(V)$ acts naturally on E and there is a bijection between the orbits of E under $GL(V)$ and the isomorphism classes of projective modules P satisfying $P \oplus R \simeq R^n$ (see [37]). Thus $GL(V)$ is transitive on E if and only if whenever P satisfies $P \oplus R \simeq R^n$ then P is free. This was relevant to Serre's Conjecture "Projective modules over $k[X_1,\ldots,X_n]$ are free (k a field)." discussed in (I.G) since if $R = k[X_1,\ldots,X_n]$ then a projective R-module P has a free complement, i.e., there exist free spaces V_0 and V_1 with $P \oplus V_0 = V_1$ and thus by induction the conjecture reduces to cancelling R in $P \oplus R \simeq R^n$.

We now comment on $Aut(GL(V))$.

In 1928, as noted earlier, Schreier and Van der Waerden [132] determined the automorphisms of $GL_n(R)$ when R was a field. Later Dieudonné [63] described the automorphisms in the case that R was a division ring.

Almost immediately, Hua and Reiner [81] determined the automorphisms when R was the ring of rational integers. Reiner, together with Landin, in 1957 [98] extended the results to non-commutative principal ideal domains. In the late 60's the automorphisms of $GL_n(R)$ over integral domains were examined separately by O'Meara and Yen. Yen [155] obtained partial results while O'Meara [108] was able to characterize completely the automorphisms of $GL_n(R)$ over <u>any</u> commutative domain R of <u>any</u> characteristic when $n \geq 3$! In each of the above cases, the automorphisms appear in the form described in this chapter. In 1972, we [116] showed the automorphisms of $GL_n(R)$, where R was a local ring, have the desired form. Keenan [88] in an unpublished thesis in 1965 also determined the automorphisms of $GL_n(R)$ when R was a local ring. Keenan's thesis has escaped notice until recently. Indeed, part of the motivation for this monograph was to illustrate the results of Keenan.

O. T. O'Meara has been principally responsible for renewed work in the problem of characterizing the automorphisms of the classical linear groups. His method of residual spaces, in the domain case, is beautiful and encompasses a far greater variety of groups than here-to-now have been able to be handled. An excellent exposition of these techniques for $GL_n(R)$ is available in [23]. It is unfortunate that at the time of this writing the O'Meara approach has not been extended to rings possessing zero divisors.

Historically, four approaches have developed.

(a) The method of involutions by Rickart and Dieudonné. This is the approach used in this chapter.

(b) The method of residual spaces. This is the O'Meara School. See [23].

(c) The method of Borel and Tits. This is very recent, employs unipotent transformations, and appears in [46]. This develops a homomorphism and isomorphism theory for a wide class of groups using the theory of algebraic groups. Their results are related to linear groups and groups having non-trivial hyperbolic rank (see III - V).

(d) The matrix method or the Chinese School [155].

The first three approaches above use the given automorphism to establish a projectivity and then employ the Fundamental Theorem of Projective Geometry (FTPG) to create a semi-linear isomorphism which gives rise to the "inner" portion of the automorphism. The approach of the Chinese School is matrix theoretic and does not invoke FTPG. (Indeed, the arguments appear to reprove this result in each case. However, it is difficult to identify where the theorem actually appears in the proofs.) However, the advantage of this highly computational argument is that in dealing with matrices and their elements, it is easier to allow a greater variety of choices of a scalar ring, e.g., allow zero divisors. This motivated the work in [101].

The above arguments for the most part assume $\dim(V) \geq 3$ and 2 is a unit in R . The case dim(V) = 2 is pathological. Reiner [118] has shown a new class of automorphisms appear for $GL_2(R)$, $R = k[X]$ for k a field, and recent work by Dull [65] shows that the appearance of a new class of automorphisms is a function of the size of the group R^* of units of R . A striking result by Cohn [55] is that if A is any free associative algebra over a commutative field on at most countably many free generators then $GL_2(A)$ is isomorphic to $GL_2(k[X])$. A remark here is necessary. In general, the automorphism problem for $GL_n(R)$ is stated as follows:

If $\Lambda : GL_n(R) \to GL_m(S)$ is a group isomorphism, does this imply

(a) $R \simeq S$ (as rings),

(b) $n = m$,

and

(c) Λ has the form of the automorphisms in this chapter?

Ideally, one would like to extend the residual space approach of O'Meara to commutative rings. This might be approached as follows: Define a pair $\langle R,S \rangle$ (where R is a commutative ring and S a multiplicative subset of R containing no zero divisors) to be "stable" if the ring of fractions $S^{-1}R$ has all projective modules free. This setting could be used to parallel the interplay between the domain and its quotient field of O'Meara's approach. Utilizing this setting, in the matrix approach [101], it was shown when $n \geq 3$ and 2 is a unit that the automorphisms of $GL_n(R)$ were of the desired form when either

(a) R is a Euclidean domain, a local ring or a connected semi-local ring and S = units of R

or

(b) R is a Dedekind domain whose quotient field is a finite extension of the rationals or a unique factorization domain with a Cooke [60] k-stage algorithm and $S = R - \{0\}$.

The assumption that 2 is a unit is crucial to any approach employing involutions. It is not necessary in the residual space approach. Indeed, drawbacks of the method of involutions is that one needs 2 a unit and that many classical linear groups do not contain a sufficient number of involutions. This was a motivation behind the residual space technique.

A good discussion of the automorphisms of $GL(V)$ when $\dim(V) = 2$ when R is a domain together with historical comments is provided by M. Dull in [65]. As noted earlier, in dimension 2 a new class of automorphisms — the Reiner automorphism — appears. This class of automorphisms does not appear in the local ring setting, however, it does occur for two dimensional general linear groups over polynomial rings over fields. These automorphisms are constructed as follows: Let k be a field and σ an automorphism of $k[X]$ <u>as a vector space over</u> k satisfying $\sigma(1) = 1$. For $A = [a_{ij}]$ in $GL_2(k[X])$, define A^σ by $A^\sigma = [\sigma a_{ij}]$. The generators of $GL_2(k[X])$ are the matrices

$$\begin{bmatrix} \alpha & 0 \\ 0 & \beta \end{bmatrix} \quad (\ \alpha,\beta \ \text{ units}) \ ,$$

$$\begin{bmatrix} 1 & 0 \\ \delta & 1 \end{bmatrix} , \begin{bmatrix} 1 & \delta \\ 0 & 1 \end{bmatrix} \quad (\ \delta \ \text{ in } \ k[X]\ ) \ .$$

The map $A \to A^{\sigma}$ applied to these generators induces an automorphism on $GL_2(k[X])$, but it is <u>not</u> the case that $A \to A^{\sigma}$ for <u>all</u> A in $GL_2(k[X])$ under this automorphism. For a further discussion of this see Dull [65] and Cohn [55].

An interesting and useful source, discussing the directions and problems in this area is given in G. Baumslag's <u>Reviews on Infinite Groups</u> on pages 775-857, in Chapter XXI, Classical Groups. We highly recommend this source. A good survey of the geometric algebra over fields containing an excellent bibliography of the literature through the early 60's is J. Dieudonné's <u>La Géométrie des Groupes Classiques</u>, (Springer-Verlag (1955), revised (1963)).

III. BILINEAR FORMS AND ISOMETRY GROUPS

(A) INTRODUCTION AND BASIC CONCEPTS

In this chapter we introduce the concept of a bilinear form on an R-space. By assuming orthogonality conditions, the form induces a geometry on the space. Our purpose is to examine the subgroup of the general linear group which leaves invariant a given form.

The initial material is based on expositions by Milnor [20] and Bass [8]. Here we discuss a bilinear space, the matrix of a form, orthogonality and a λ-inner product. The presentation is somewhat more general than Milnor's [20] but considerably less extensive than Bass's [8]. One will observe that most of the results through Section (D) could be stated for an arbitrary commutative ring (occasionally having 2 a unit).

Section (D) introduces the unitary transvection. This isometry will play a role analogous to the transvection in the general linear group and will appear repeatedly throughout the remainder of the monograph. The unitary transvection will enable us to describe the action of a group fixing a form as a transformation group on the space and, in turn, determine its transitivity on vectors, lines and planes.

Let R denote a commutative local ring. Let V and W be R-spaces of finite dimension. An R-bilinear map or pairing

$$\beta : V \times W \to R$$

is a mapping such that $\beta(v,w)$ is R-linear as a function of v for fixed w, and R-linear as a function of w for fixed v.

If $V^* = \mathrm{Hom}_R(V,R)$ denotes the dual of V, we have a natural pairing

$$(\,,)_V : V^* \times V \to R$$

given by

$$(f,v)_V = f(v) . \qquad \text{(See (I.F) or (II.F).)}$$

A morphism $\sigma : V \to W$ induces naturally $\sigma^* : W^* \to V^*$ by

$$(\sigma^* f,v)_V = (f,\sigma v)_W .$$

The morphism σ^* is called the adjoint or transpose (see the remark following (I.25)).

Let $\mathrm{Bil}(V \times W)$ denote the R-module of all R-bilinear maps $\beta : V \times W \to R$. A given bilinear map β induces two linear morphisms

$$\beta^d : V \to W^* ,$$
$$d_\beta : W \to V^*$$

given by $({}_{\beta}dv,w)_W = \beta(v,w)$ and $(d_{\beta}w,v)_V = \beta(v,w)$. We call β an <u>inner product</u> if d_{β} and ${}_{\beta}d$ are isomorphisms.

In the case of the pairing

$$(\,,\,)_V : V^* \times V \to R ,$$

denote $d_{(\,,\,)_V}$ by d_V . Observe d_V gives a natural morphism

$$d_V : V \to V^{**}$$

by

$$(d_V x, y)_{V^*} = (y,x)_V .$$

Since V is finite dimensional, d_V is an isomorphism and we identify V with V^{**} via d_V .

If $\beta : V \times W \to R$ is bilinear and v is in V and w is in W , then

$$\begin{aligned}(d_{\beta}^* \circ d_V(v),w)_W &= (d_V(v),d_{\beta}(w))_{V^*} \\ &= (d_{\beta}(w),v)_V \\ &= \beta(v,w) \\ &= ({}_{\beta}d(v),w)_W .\end{aligned}$$

That is, ${}_{\beta}d = d_{\beta}^* \circ d_V$. Similarly, $d_{\beta} = {}_{\beta}d^* \circ d_W$. Identifying V with V^{**} and W with W^{**} (thus d_W = identity and d_V = identity), we have the following lemma.

<u>(III.1)</u> <u>LEMMA</u>. (For the above setting) The following are equivalent:

(a) β is an inner product.

(b) d_β is an isomorphism.

(c) ${}_\beta d$ is an isomorphism.

A pair (V,β) where V is a finite dimensional R-space and $\beta : V \times V \to R$ is a bilinear form is called a <u>bilinear space</u>. We remark that adjectives applied to either β or V will be applied to (V,β) , e.g., if β is an inner product then (V,β) is an inner product space.

Suppose (V,β) has a basis $\{b_1,\ldots,b_n\}$. Let $\beta_{ij} = \beta(b_i,b_j)$. Then the <u>matrix of the form</u> β is $[\beta_{ij}]$. If x and y are in V , say

$$x = \sum \alpha_i b_i$$

$$y = \sum \delta_i b_i ,$$

then

$$\beta(x,y) = \sum_i \sum_j \alpha_i \delta_j \beta(b_i,b_j)$$

$$= \sum_i \sum_j \alpha_i \delta_j \beta_{ij} .$$

Letting $B = [\beta_{ij}]$ and identifying x with $\langle\alpha_1,\ldots,\alpha_n\rangle^t$ and y with $\langle\delta_1,\ldots,\delta_n\rangle^t$, we have

$$\beta(x,y) = x^t B y .$$

Conversely, if $B = [h_{ij}]$ is in $(R)_n$ then define $\beta : V \times V \to R$ by defining $\beta(b_i, b_j) = h_{ij}$ and extending linearly. This gives a natural bijection between $\mathrm{Bil}(V \times V)$ and $(R)_n$.

Notation: If (V,β) has matrix $B = [\beta_{ij}] = [\beta(b_i, b_j)]$ for a basis $\{b_1, \ldots, b_n\}$ then we denote (V,β) by $(V, \langle B \rangle)$ or $\langle B \rangle$.

(III.2) THEOREM. Let (V,β) be a bilinear space. Let B denote the matrix of β relative to a fixed basis. The following are equivalent:

(a) β is an inner product.

(b) B is invertible.

(c) $\det(B)$ is a unit.

Proof. Let V have basis $b_1, \ldots, b_n$ where $B = [\beta_{ij}] = [\beta(b_i, b_j)]$. Let V^* have dual basis $b_1^*, \ldots, b_n^*$ given by $(b_i^*, b_j)_V = \delta_{ij}$. Then the map

$$\beta^d : V \to V^*$$

given by $\beta^d(v) = \beta(v, \ )$ is given by $b_i \to \beta(b_i, \ ) = \sum_j \beta_{ij} b_j^*$. Thus, the map is bijective if and only if B is invertible.

What happens to B under a change of basis? Let (V,β) and (W,σ) be bilinear spaces. A morphism

$$\alpha : (V,\beta) \to (W,\sigma)$$

is an R-linear morphism $\alpha : V \to W$ satisfying $\sigma(\alpha x, \alpha y) = \beta(x,y)$ for all x and y in V. If α is bijective then α is called an isometry and V and W are called isometric, written $V \simeq W$.

Let

$$U(V) = \{\alpha : V \to V \mid \alpha \text{ is an isometry}\}$$
$$= \{\alpha \in GL(V) \mid \beta(x,y) = \beta(\alpha x, \alpha y)\}$$

denote the group of isometries of V.

(III.3) THEOREM. Let V be an R-space. Suppose V has bilinear forms $\langle B \rangle$ and $\langle \bar{B} \rangle$. Then $(V, \langle B \rangle) \simeq (V, \langle \bar{B} \rangle)$ if and only if there is an invertible matrix A with $ABA^t = \bar{B}$.

Proof. Let $\{b_1, \ldots, b_n\}$ be a basis for V giving rise to B. If $\{\bar{b}_1, \ldots, \bar{b}_n\}$ is another basis then

$$\bar{b}_i = \sum_{j=1}^{n} \alpha_{ji} b_j .$$

Further, $[\alpha_{ji}]$ is invertible. Thus, if the matrix of $B = [\beta_{ij}] = [\beta(b_i, b_j)]$ for $\beta : V \times V \to R$ then

$$\beta(\bar{b}_i, \bar{b}_j) = \sum_{k,\ell} \alpha_{ik} \beta_{k\ell} \alpha_{j\ell} .$$

The conclusion follows.

Example: Let (V,β) be a bilinear space $V = Re$ of dimension 1 where $\beta(x,y) = \beta(y,x)$ for x,y in V. Then $B = [\beta(e,e)] = [v]$. Denote (V,β) by $\langle v\rangle$. If β is an inner product then v is a unit. Further, $\langle v\rangle \simeq \langle\bar{v}\rangle$ if and only if $v = \alpha^2\bar{v}$ where α is a unit, i.e., $v \equiv \bar{v} \mod (R^*)^2$

Although we may continue a discussion of bilinear forms, the sharpest results begin to appear when we impose 'orthogonality' conditions on an inner product. By this we mean the following: If (V,β) is an inner product space and x and y are in V it is natural to say that x is orthogonal to y if $\beta(x,y) = 0$. However, we would like this to be reflexive, i.e., if $\beta(x,y) = 0$ then $\beta(y,x) = 0$.

To assure this we introduce a "λ-inner product."

Let λ be an element of R satisfying $\lambda^2 = 1$. A form β in $Bil(V \times V)$ is called λ-<u>bilinear</u> if $\beta(x,y) = \lambda\beta(y,x)$.

If 2 is a unit in R then $X^2 - 1$ is a separable polynomial and has only 1 and -1 as its zeros. Then $\lambda = 1$ or $\lambda = -1$.

Suppose β is a λ-bilinear form. If $\lambda = 1$, then β is called <u>symmetric</u> and

$$\beta(x,y) = \beta(y,x) .$$

If $\lambda = -1$, then β is called <u>skew-symmetric</u> and

$$\beta(x,y) = -\beta(y,x) .$$

If a form β satisfies $\beta(x,x) = 0$ for all x in V then β is _symplectic_. Observe every symplectic inner product space is skew-symmetric since

$$\beta(x,y) + \beta(y,x) = \beta(x + y,x + y) - \beta(x,x) - \beta(y,y) = 0 .$$

Conversely, if 2 is a unit in R, then skew-symmetric inner products are symplectic since $\beta(x,x) = -\beta(x,x)$ implies $\beta(x,x) = 0$.

Let (V,β) be a λ-bilinear space. Let x and y be in V. Then x is _orthogonal_ to y if $\beta(x,y) = 0$, written $x \perp y$.

Suppose (V,β) is a λ-inner product space and 2 is a unit of R. Let $\{b_1,\ldots,b_n\}$ be a basis for V and $B = [\beta_{ij}]$ be the matrix of β relative to this basis. If $\lambda = 1$ then $\beta_{ij} = \beta_{ji}$, i.e., $B^t = B$, and B is symmetric. Conversely, symmetric invertible matrices give rise to 1-inner products. If $\lambda = -1$ then $\beta_{ij} = -\beta_{ji}$ and $B^t = -B$, that is, B is a skew-symmetric matrix. Conversely, skew-symmetric invertible matrices give rise to skew-symmetric inner products.

Continuing, if A is in $(R)_n$ set

$$A_1 = \frac{1}{2}(A + A^t)$$

$$A_2 = \frac{1}{2}(A - A^t) .$$

Then A_1 is symmetric, A_2 is skew-symmetric and $A = A_1 + A_2$.

Denote $\text{Bil}(V \times V)$ by $\text{Bil}(V)$ and let λ be in R with $\lambda^2 = 1$. For β in $\text{Bil}(V)$ define β^* in $\text{Bil}(V)$ by

$$\beta^*(x,y) = \beta(y,x) .$$

Then $\text{Bil}(V)$ is an R-module with an R-linear involution $\beta \to \beta^*$.

The map $\beta \to d_\beta$ from $\text{Bil}(V)$ to $\text{Hom}_R(V,V^*)$ satisfies $\beta^* \to d_{\beta^*} = (d_\beta)^*$ when we identify $V \equiv V^{**}$ and is thus an involution preserving R-isomorphism.

The map $\beta \to \lambda\beta^*$ is an R-linear automorphism of order ≤ 2. If we define

$$S_\lambda : \text{Bil}(V) \to \text{Bil}(V)$$

by

$$S_\lambda(\beta) = \beta + \lambda\beta^*$$

then S_λ is an R-morphism and

$$\text{Ker}(S_{-\lambda})$$

is precisely the module of λ-bilinear forms on V.

(B) λ-INNER PRODUCT SPACES

Let R be a local commutative ring and λ be in R with $\lambda^2 = 1$.

Suppose $(V_1,\beta_1),\ldots,(V_t,\beta_t)$ are λ-bilinear spaces. Then (V,β) is the orthogonal sum of $\{(V_i,\beta_i)\}$, denoted

$$V = V_1 \perp \cdots \perp V_t ,$$

if

(a) $V = V_1 \oplus \cdots \oplus V_t$;

(b) β is given by

$$\beta(v_1 + \cdots + v_t, w_1 + \cdots + w_t) = \textstyle\sum_{i=1}^{t} \beta_i(v_i,w_i)$$

where v_i and w_i are in V_i , $1 \le i \le t$.

We write β as $\beta = \beta_1 \perp \cdots \perp \beta_t$.

Let (V,β) be a λ-bilinear space and W a submodule of V . Then the orthogonal complement of W ,

$$\begin{aligned} W^{\perp} &= \{v \text{ in } V \mid \beta(w,v) = 0 \text{ for all } w \text{ in } W\} \\ &= \{v \text{ in } V \mid \beta(W,v) = 0\} . \end{aligned}$$

Note $W^{\perp}$ may not be a "complement" of W in the usual sense. The following is immediate.

(III.4) LEMMA. Let (V,β) be a λ-bilinear space and W a submodule of V. Then

(a) $W^{\perp}$ is a submodule of V,

(b) if $W \subset P$, P a submodule, then $P^{\perp} \subset W^{\perp}$,

(c) $W \subset W^{\perp\perp}$,

(d) $W^{\perp} = W^{\perp\perp\perp}$.

Suppose (V,β) is a λ-inner product and W is a subspace. Though β is an inner product its restriction $\beta|_{W\times W}$ is a bilinear form but may not be an inner product. If the restriction is an inner product then W splits V.

(III.5) THEOREM. (Orthogonal Decomposition Lemma) Let $\sigma : (V,\beta) \to (W,\alpha)$ be a morphism of λ-bilinear spaces. Suppose β is an inner product. Then σ is injective and

$$W = \sigma V \perp (\sigma V)^{\perp} .$$

Proof. If $\sigma v = 0$ then for all w in V, $\beta(v,w) = \alpha(\sigma v,\sigma w) = 0$. Since β is an inner product, $v = 0$ and σ is injective. Since σ is injective, we may identify V with a submodule of W so that $\beta = \alpha|_{V\times V}$. Let w be in W. The linear form $v \to \alpha(w,v)$ on V may be represented by $v \to \beta(y,v)$ for a unique y in V since β is an inner product. Then $w = y + (w - y)$ where y is in V and $w - y$ is in $V^{\perp}$.

(III.6) COROLLARY. Let (V,β) be a λ-bilinear space. Let $\{b_1,\ldots,b_t\}$ be elements of V such that $[\beta(b_i,b_j)]$ is invertible. Then

(a) $\{b_1,\ldots,b_t\}$ is R-free and extends to a basis of V.

(b) If $W = \oplus \sum Rb_i$ then $\beta|_{W\times W}$ is an inner product and $V = W \perp W^{\perp}$.

Proof. Let W be the submodule generated by $\{b_1,\ldots,b_t\}$. The set $\{b_1,\ldots,b_t\}$ is R-free since if $\alpha_1 b_1 + \cdots + \alpha_t b_t = 0$ for non-trivial α_i then $[\beta(b_i,b_j)]$ would not be invertible. Then $(W,\beta|_{W\times W})$ is an inner product space. Now apply the above theorem.

We often apply (III.6) in the following fashion when $\lambda = 1$. Let x be in (V,β) and suppose $\beta(x,x) = u$ where u is a unit. Then x is unimodular and $\beta|_{Rx\times Rx}$ is an inner product. Further
$V = Rx \perp (Rx)^{\perp} \simeq \langle u\rangle \perp W$, $W = (Rx)^{\perp}$.

(III.7) COROLLARY. Let (V,β) be a 1-bilinear space. Then

$$V \simeq \langle u_1\rangle \perp \cdots \perp \langle u_s\rangle \perp W \qquad (s \geq 0)$$

where the u_i are units and $\beta(x,x)$ is a non-unit for all x in W. Further, if 2 is a unit and β is an inner product, then $W = 0$.[1]

(III.8) LEMMA. Let (V,β) be a λ-inner product space. Let W be a subspace of V. Then

(a) $W^{\perp}$ is a subspace of V.

(b) $W = W^{\perp\perp}$.

---

[1] If $W \neq 0$ select a basis $\{b_1,\ldots,b_t\}$ for W. Then select $\{c_1,\ldots,c_t\}$ in W with $\beta(c_i,b_j) = \delta_{ij}$. Then $2 = 2\beta(c_1,b_1) = \beta(c_1 + b_1,c_1 + b_1) - \beta(c_1,c_1) - \beta(b_1,b_1)$ and hence 2 is a non-unit - a contradiction.

(c) $W^{\perp} \simeq (V/W)^{*}$ and $W^{*} \simeq V/W^{\perp}$.

<u>Proof</u>. Since $0 \to W \to V \to V/W \to 0$ splits, so does

$$0 \to (V/W)^{*} \to V^{*} \to W^{*} \to 0 \ .$$

We have from β the following diagram:

$$\begin{array}{ccccccccc} 0 & \longrightarrow & (V/W)^{*} & \longrightarrow & V^{*} & \xrightarrow{r} & W^{*} & \longrightarrow & 0 \\ & & & & \uparrow{\scriptstyle \beta^{d}} & & & & \\ 0 & \longrightarrow & W & \longrightarrow & V & \longrightarrow & V/W & \longrightarrow & 0 \end{array} \ .$$

Observe
$$\begin{aligned} {}_{\beta}d^{-1}((V/W)^{*}) &= \operatorname{Ker}(r_{\beta}d) \\ &= \{v \text{ in } V \mid \beta(v,w) = 0 \text{ for all } w \text{ in } W\} \\ &= W^{\perp} \ . \end{aligned}$$

Thus

$$0 \longrightarrow W^{\perp} \longrightarrow V \xrightarrow{r_{\beta}d} W^{*} \longrightarrow 0$$

is exact. Hence $W^{*} \simeq V/W^{\perp}$, $W^{\perp} = {}_{\beta}d^{-1}((V/W)^{*})$ and $W^{\perp}$ is a direct summand.

Let (V,β) be a λ-bilinear space. If W is a subspace of V and $\beta|_{W\times W}$ is an inner product, then W is called a <u>non-singular subspace</u>. The opposite of non-singularity would be $\beta|_{W\times W} = 0$, i.e., $W \subset W^{\perp}$, in this case W is <u>totally isotropic</u>. The relation between maximal totally isotropic subspaces and the hyperbolic spaces of the next section is given in (III.11).

(C) HYPERBOLIC PLANES

Let (V,β) be a λ-bilinear space. The <u>hyperbolic space of</u> V <u>relative to</u> β, denoted

$$H(V,\hat{\beta}) ,$$

is the space $V \oplus V^*$ where

$$\hat{\beta}(v + f,w + g) = f(w) + \lambda g(v) + \beta(v,w)$$

where v and w are in V, f and g are in V^*.

The morphism

$$d_{\hat{\beta}} : V \oplus V^* \to V^* \oplus V = (V \oplus V^*)^*$$

is represented by the matrix

$$\begin{bmatrix} \beta^d & 1 \\ \lambda d_V & 0 \end{bmatrix}$$

where d_V was given in Section (III.A) and is invertible.

If V is an R-space, then $H(V,0)$ is the space $V \oplus V^*$ with inner product

$$0(v + f,w + g) = f(w) + \lambda g(v)$$

(III.9) THEOREM. Let (V,β) and (W,α) be λ-bilinear spaces. Then

(a) $H(V,\hat{\beta})$ is a λ-inner product space.

(b) If 2 is a unit in R, then

$$H(V,\hat{\beta}) \simeq H(V,0) .$$

(c) If $\sigma : (V,\beta) \to (W,\alpha)$ is an isometry, then $H(\sigma) = \sigma \oplus \sigma^{*-1}$ is an isometry $H(V,\hat{\beta}) \to H(W,\hat{\alpha})$.[1]

(d) There is a natural isometry

$$H(V,\hat{\beta}) \perp H(W,\hat{\alpha}) \to H(V \perp W,\widehat{\beta \perp \alpha})$$

given by

$$(v + f) + (w + g) \to (v + w) + (f + g)$$

where $v \in V$, $f \in V^*$, $w \in W$, $g \in W^*$.

Proof. To show (b) we show a more general result. Let σ be a bilinear form on V. We show

$$\sum = \begin{bmatrix} 1 & 0 \\ \sigma^d & 1 \end{bmatrix} : V \oplus V^* \to V \oplus V^*$$

is an isometry from $H(V,\hat{\beta})$ to $H(V,\widehat{\beta - 2\sigma})$. (Observe, if $\bar{\sigma}$ is defined by $\bar{\sigma}(x,y) = \sigma(y,x)$, then $2\sigma = \sigma + \lambda\bar{\sigma}$ and 2σ is λ-bilinear. Thus, since 2 is a unit, $\beta = 2\sigma$ where $\sigma = \frac{1}{2}\beta$ and (b) follows as a special case of the above statement.)

---

[1] Observe $H : (V,\beta) \to H(V,\hat{\beta})$ is a "functor."

the forms. Set $\Lambda = \widehat{\beta - 2\sigma}$. Then, for v,w in V and f,g in V^* ,

$$\begin{aligned}\Lambda(\textstyle\sum(v + f),\sum(w + g)) &= \Lambda(v + (f + {}_\sigma dv), w + (g + {}_\sigma dw)) \\ &= (f + {}_\sigma dv)(w) + \lambda(g + {}_\sigma dw)(v) + (\beta - 2\sigma)(v,w) \\ &= f(w) + \sigma(v,w) + \lambda g(v) + \lambda\sigma(w,v) + \beta(v,w) - 2\sigma(v,w) \\ &= f(w) + \lambda g(v) + \beta(v,w) \\ &= \hat{\beta}(v + f, w + g) .\end{aligned}$$

This gives (b). Parts (c) and (d) are verified in a similar fashion.

Throughout these notes we will be principally concerned with the case where 2 is a unit of R . Thus, by (b) above, we may take the hyperbolic space $H(V,\hat{\beta})$ of (V,β) to be of the form $H(V,0)$.

Denote $H(V,0)$ by $H(V)$ and denote the bilinear form on $H(V)$ by h .[1] Thus

$$H(V) = V \oplus V^*$$

and

$$h(v + f, w + g) = f(w) + \lambda g(v)$$

for v,w in V and f,g in V^* . Let $\{b_1,\ldots,b_n\}$ be a basis of V and $\{b_1^*,\ldots,b_n^*\}$ a dual basis for V^* . Then, the union of these bases is a basis for $H(V)$ and the matrix of h relative to this basis is

---

[1] In this case, $H(\ )$ is referred to as the <u>hyperbolic functor</u> which carries R-spaces to hyperbolic spaces.

$$\begin{bmatrix} 0 & \lambda I \\ I & 0 \end{bmatrix} .$$

(III.10) PROPOSITION. Let (V,β) be a λ-inner product space. Then

$$(V,\beta) \perp (V,-\beta) \simeq H(V,\hat{\beta})$$

under

$$\sigma : v + w \to (v - w) + {}_{\beta}dw$$

where v,w are in V and ${}_{\beta}dw$ is in V^*. Hence, if 2 is a unit in R by (III.9)(b)

$$(V,\beta) \perp (V,-\beta) \simeq H(V) .$$

Proof. Observe $\mathrm{Im}(\sigma) = V \oplus \mathrm{Im}({}_{\beta}d) = V \oplus V^*$ since β is an inner product. Thus σ is an isomorphism of R-spaces. It remains to show σ preserves the forms. Let v,w be in (V,β) and x,y be in $(V,-\beta)$. Then

$$\begin{aligned}
&\hat{\beta}(\sigma(v + x),\sigma(w + y)) \\
&\quad = \hat{\beta}(v - x + {}_{\beta}dx, w - y + {}_{\beta}dy) \\
&\quad = {}_{\beta}dx(w - y) + \lambda\, {}_{\beta}dy(v - x) + \beta(v - x, w - y) \\
&\quad = \beta(x, w - y) + \lambda\beta(y, v - x) + \beta(v - x, w - y) \\
&\quad = \beta(v - x, y) + \beta(v, w - y) \\
&\quad = \beta(v,w) - \beta(x,y) \\
&\quad = (\beta \perp (-\beta))(v + x, w + y) .
\end{aligned}$$

Let (V,β) be a λ-inner product space. Then (V,β) is <u>split</u> if V has a subspace W with $W^{\perp} = W$, i.e., W is a maximal totally isotropic subspace.

<u>(III.11)</u> <u>THEOREM</u>. Let (V,β) be a λ-inner product space. Then the following are equivalent:

(a) (V,β) is split.

(b) There is a subspace W of V with $(V,\beta) \simeq H(W,\hat{\beta})$.

<u>Proof</u>. If $(V,\beta) = H(W,\hat{\beta})$, then $V = W \oplus W^*$ and $(W^*)^{\perp} = W^*$. Conversely, if V is split and $V = W \oplus U$ where $U^{\perp} = U$ then it is easy to check that $u \to {}_{\beta}du|_W$ defines an isomorphism $U \to W^*$ and $W \oplus U \to H(W,\hat{\beta})$ by $w + u \to w + {}_{\beta}du|_W$ is an isometry.

Let (V,β) be a split λ-inner product space. If $\dim(V) = 2$ then (V,β) is called a <u>hyperbolic plane</u>.

A λ-inner product space (V,β) is said to have <u>hyperbolic rank greater than or equal to</u> t if

$$V = H_1 \perp \cdots \perp H_t \perp W$$

where H_i are hyperbolic planes.

<u>(III.12)</u> <u>LEMMA</u>. Let (V,β) be a λ-inner product space of $\dim(V) = 2$. Let 2 be a unit in R. Then, the following are equivalent:

(a) (V,β) is a hyperbolic plane.

(b) $(V,\beta) \simeq (V, \begin{bmatrix} 0 & \lambda \\ 1 & 0 \end{bmatrix})$.

Further, if $\lambda = 1$ then the above are equivalent to

(c) $(V,\beta) \simeq \langle 1 \rangle \perp \langle -1 \rangle$.

Proof. The fact that (a) and (b) are equivalent follows from (III.11) and (III.9). To show (b) is equivalent to (c) when $\lambda = 1$, note

$$\begin{bmatrix} 1 & -1 \\ \frac{1}{2} & \frac{1}{2} \end{bmatrix}^t \begin{bmatrix} 0 & 1 \\ 1 & 0 \end{bmatrix} \begin{bmatrix} 1 & -1 \\ \frac{1}{2} & \frac{1}{2} \end{bmatrix} = \begin{bmatrix} 1 & 0 \\ 0 & -1 \end{bmatrix} .$$

(III.13) PROPOSITION. Let (V,β) be a split λ-inner product space. Let 2 be a unit of R . Then $(V,\beta) = H_1 \perp \cdots \perp H_t$ where H_i are hyperbolic planes.

Proof. This follows from the discussion after (III.9).

(III.14) LEMMA. Let (V,β) be a λ-inner product space. Let e in V be unimodular and $\beta(e,e) = 0$. Then there is an f in V satisfying

(a) $H = Re \oplus Rf$ is a hyperbolic plane.

(b) $\beta(f,e) = 1$.

(c) $V = H \perp W$.

An e satisfying the above is called isotropic.

Proof. Since e is unimodular, $V = Re \oplus W$. Let $e_2,\ldots,e_n$ be a basis for W , then $\{e = e_1,e_2,\ldots,e_n\}$ is a basis for V . Let $\sigma_1,\ldots,\sigma_n$ be a dual basis, $\sigma_i(e_j) = \delta_{ij}$, for V^* . Since β is an inner product there are $f = f_1, f_2, \ldots, f_n$ in V with $\sigma_i(e_j) = \beta(e_j, f_i)$. Then $\{e,f\}$ is a hyperbolic plane.

(D) TRANSVECTIONS AND THE GROUP $U(V)$

——— o o o ———

Throughout this section we assume that R is a commutative local ring having 2 a unit. Further, we assume V is a λ-inner product space with hyperbolic rank ≥ 1 . Thus, let

$$V = H \perp W$$

where H is a hyperbolic plane with

$$H = Ru \oplus Rv \;,$$

$$\beta(u,v) = 1 \;, \quad \beta(v,u) = \lambda \;,$$

$$\text{and} \quad \beta(u,u) = \beta(v,v) = 0 \;.$$

Let $\dim(V) = n \geq 2$.

——— o o o ———

Since 2 is a unit, $\lambda = 1$ or $\lambda = -1$. If $\lambda = 1$, the geometry associated with (V,β) is called <u>orthogonal</u> and the group $U(V)$ of isometries of V is called the <u>orthogonal group</u> and denoted $O(V)$. If $\lambda = -1$, the geometry associated with (V,β) is called <u>symplectic</u>

and $U(V)$ is called the <u>symplectic group</u> and denoted $Sp(V)$.

We describe some elements of $U(V)$.

(a) Let x be in V and satisfy $\beta(x,u) = 0$. Define

$$\sigma_{u,x} : V \to V$$

by

$$\sigma_{u,x}(z) = z + \beta(x,z)u - \lambda\beta(u,z)x - \frac{1}{2}\beta(x,x)\beta(u,z)u .$$

Define, for x satisfying $\beta(x,v) = 0$, the map $\sigma_{v,x}$ in a similar fashion. The maps $\sigma_{u,x}$ and $\sigma_{v,x}$ are called <u>unitary transvections</u>.[1] Often x will be taken from W .

(b) Define

$$\Delta : V \to V$$

by

$$\Delta(u) = v$$

$$\Delta(v) = \lambda u$$

and $$\Delta(w) = w \quad \text{for all } w \text{ in } W .$$

(c) Let ε be a unit of R . Define

$$\Phi_\varepsilon : V \to V$$

by

$$\Phi_\varepsilon(u) = \varepsilon u$$

$$\Phi_\varepsilon(v) = \varepsilon^{-1}v$$

and $$\Phi_\varepsilon(w) = w \quad \text{for all } w \text{ in } W .$$

---

[1] The choice of terminology is due to Bass [8]. Perhaps a better choice would be <u>Eichler transformations</u> or <u>hyperbolic plane rotations</u>.

<u>(III.15)</u> <u>PROPOSITION</u>. The maps $\sigma_{u,x}$, $\sigma_{v,x}$, Δ and Φ_ε are isometries of V .

<u>Proof</u>. It is straightforward to check that Δ and Φ_ε are isometries. Let $\sigma = \sigma_{u,x}$ and let z and $\bar{z}$ be elements of V . Consider

$$\beta(\sigma z,\sigma\bar{z}) - \beta(z,\bar{z}) = \alpha + \eta + \gamma$$

where

$$\alpha = \beta(z,\sigma\bar{z} - \bar{z})$$

$$\eta = \beta(\sigma z - z,\bar{z})$$

$$\gamma = \beta(\sigma z - z,\sigma\bar{z} - \bar{z}) .$$

We must show $\alpha + \eta + \gamma = 0$. Computing directly from the definition of $\sigma_{u,x}$:

$$\alpha = \beta(x,\bar{z})\beta(z,u) - \lambda\beta(u,\bar{z})\beta(z,x) - \frac{1}{2}\beta(x,x)\beta(u,\bar{z})\beta(z,u)$$

$$\eta = \beta(x,z)\beta(u,\bar{z}) - \lambda\beta(u,z)\beta(x,\bar{z}) - \frac{1}{2}\beta(x,x)\beta(u,z)\beta(u,\bar{z})$$

$$\gamma = \beta(u,z)\beta(u,\bar{z})\beta(x,x) .$$

Then, if $\lambda = 1$ or $\lambda = -1$, $\alpha + \eta + \gamma = 0$.

A similar verification gives the following result.

<u>(III.16)</u> <u>THEOREM</u>. For the above isometries

(a) $\sigma_{u,x}\sigma_{u,y} = \sigma_{u,x+y}$. Hence $(\sigma_{u,x})^{-1} = \sigma_{u,-x}$. Further $\sigma_{\alpha u,x} = \sigma_{u,\alpha x}$ for α a unit.

(b) $\Delta^{-1}\Phi_\varepsilon\Delta = \Phi_{\varepsilon^{-1}}$, $\Phi_\varepsilon^{-1} = \Phi_{\varepsilon^{-1}}$.

(c) $\Phi_\varepsilon\Delta\Phi_\varepsilon = \Delta$, $\Delta^4 = I$. (If $\lambda = 1$, then $\Delta^2 = I$.)

(d) If z is in V then $z = \alpha u + \delta v + y$ where y is in W, α and δ are in R. If x is in W then

$$\sigma_{u,x}(z) = [\alpha + \beta(x,y) - \frac{1}{2}\,\delta\beta(x,x)]u + \delta v + (y - \lambda\delta x) .$$

(e) If Θ is in $U(V)$ and $\Theta|_H = \mathrm{id}$, then $\Theta\sigma_{u,x}\Theta^{-1} = \sigma_{u,\Theta x}$. In general, $\Theta\sigma_{u,x}\Theta^{-1} = \sigma_{\Theta u,\Theta x}$ for Θ in $U(V)$.

(f) $\Phi_\varepsilon\sigma_{u,x}\Phi_\varepsilon^{-1} = \sigma_{u,\varepsilon x} = \sigma_{\varepsilon u,x}$ if x is in W.

Statements similar to (III.16) are also available for $\sigma_{v,x}$. For example, if x is in W, then

$$\Phi_\varepsilon\sigma_{v,x}\Phi_\varepsilon^{-1} = \sigma_{v,\varepsilon^{-1}x} .$$

and if $z = \alpha u + \delta v + y$ then (when x is in W)

$$\sigma_{v,x}(z) = \alpha u + [\delta + \beta(x,y) - \frac{1}{2}\,\alpha\lambda\beta(x,x)]v + (y - \alpha x) .$$

Suppose $z = \alpha u + \delta v + y$ is unimodular, i.e., $O(z) = R$. Then either α is a unit, δ is a unit or $O(y) = R$. Suppose α and δ are not units. Then $O(y) = R$ and $\beta(\ ,y) : W \to R$ is surjective (since $V = H \perp W$ we have $\beta|_{W\times W}$ an inner product). Thus, there is an x in W with $\beta(x,y)$ a unit in R and, consequently,

$$\alpha + \beta(x,y) - \frac{1}{2}\,\delta\beta(x,x)$$

and

$$\delta + \beta(x,y) - \frac{1}{2}\alpha\lambda\beta(x,x)$$

are units. Thus, a suitable transformation of z by $\sigma_{u,x}$ or $\sigma_{v,x}$ will produce a new element having either the coefficient of u or the coefficient of v, respectively, a unit of R.

Thus, without loss of generality, assume

$$z = \alpha u + \delta v + y$$

is unimodular where α or δ is a unit. Suppose δ is a unit. Set $x = \delta^{-1}\lambda y$. Then,

$$\sigma_{u,x}(z) = [\alpha + \beta(x,y) - \frac{1}{2}\delta\beta(x,x)]u + \delta v .$$

If α is a unit, then set $x = \alpha^{-1}y$ and

$$\sigma_{v,x}(z) = \alpha u + [\delta + \beta(x,y) - \frac{1}{2}\alpha\lambda\beta(x,x)]v .$$

Since $\sigma_{u,x}$ and $\sigma_{v,x}$ are in $GL(V)$, $\sigma_{u,x}(z)$ and $\sigma_{v,x}(z)$ remain unimodular.

We conclude from the above discussion that given unimodular z in V, there is an element σ in $U(V)$ with $\sigma(z) \in H$.

Suppose $x = \alpha u + \delta v$ is unimodular in H. Either α or δ is a unit. If δ is a unit, $\Delta x = \lambda\delta u + \alpha v$. Thus, without loss, we may assume α is a unit in the above form for x. Then, applying $\Phi_{\alpha^{-1}}$

to x, we may assume

$$x = u + \eta v \quad (\eta \text{ in } R).$$

Case (1). $\lambda = 1$. Suppose y is unimodular in H with $\beta(y,y) = \beta(x,x)$. We may assume y has been transformed by the above discussion into the vector $u + \bar{\eta}v$. Then $2\eta = \beta(x,x) = \beta(y,y) = 2\bar{\eta}$. Since 2 is a unit $\eta = \bar{\eta}$.

Case (2). $\lambda = -1$. Let $x = u + \eta v$. For w unimodular in H define $\tau_w : H \to H$ by $\tau_w(z) = z - \beta(w,z)w$. It is easy to check that τ_w is in $U(H)$. The element τ_w is a symplectic transvection. Observe

$$\tau_w = \sigma_{w,-w/2}$$

thus τ_w is a special case of a unitary transvection. Then, if $w = v + x$,

$$\tau_{v+x}(v) = v - \beta(v + x,v)(v + x) = -x .$$

Hence $-\tau_{v-x}^{-1}(x) = v$.

(III.17) THEOREM. The group $U(V)$ is transitive on unimodular vectors in V of the same norm.

Proof. Let x and y be in $V = H \perp W$. Then by the above discussion there exist σ_1 and σ_2 in $U(V)$ with

$$\sigma_1 x = \sigma_2 y = v \qquad \text{if } \lambda = -1 ,$$

$$\sigma_1 x = \sigma_2 y = u + \eta v \qquad \text{if } \lambda = 1 \quad \text{where } 2\eta = \beta(x,x) = \beta(y,y) .$$

Hence $\sigma_2^{-1}\sigma_1 x = y$.

(III.18) COROLLARY. The group $U(V)$ is transitive on hyperbolic planes in V .

Proof. Let $V = H \perp W$ where $H = Ru \oplus Rv$ and $\beta(u,v) = 1$. Let $\{y,x\}$ be a second hyperbolic pair in V , i.e., $\beta(y,y) = \beta(x,x) = 0$ and $\beta(y,x) = 1$. Transforming $\{y,x\}$ by a suitable element of $U(V)$, we may assume $u = y$. Then

$$x = \alpha u + \delta v + w \qquad \text{where } w \text{ is in } W .$$

Since $1 = \beta(u,x) = \beta(y,x)$, $x = \alpha u + v + w$. A suitable $\sigma = \sigma_{u,z}$ will carry x into H and simultaneously fix u .

Thus, we have

$$\begin{Bmatrix} u = y \\ x \end{Bmatrix} \xrightarrow{\sigma} \begin{Bmatrix} u \\ \alpha u + v \end{Bmatrix} \xrightarrow{\tau} \begin{Bmatrix} u \\ v \end{Bmatrix}$$

where $\tau = I_H$ if $\lambda = 1$ and τ is a suitable symplectic transvection if $\lambda = -1$.

(III.19) COROLLARY. (Cancellation) Let U and Y be λ-inner product spaces. If $U \perp V \simeq U \perp Y$ then $V \simeq Y$.

Proof. It suffices to prove the result when $U = \bar{H}$ where $\bar{H}$ is a hyperbolic plane. This is clear since if $U = (U,\sigma)$ then U may be replaced by $(U,\sigma) \perp (U,-\sigma)$ which by (III.10) is a direct sum of hyperbolic planes. Then, by induction we may assume $U = \bar{H}$.

Let $\sigma : \bar{H} \perp V \xrightarrow{\simeq} \bar{H} \perp Y$. Identify $\bar{H} \perp V$ with $\bar{H} \perp Y$ via the isometry σ and denote the image of $\bar{H}$ by $\bar{H}'$ and the image of V by V. If $\{x,y\}$ and $\{x',y'\}$ are hyperbolic pairs generating the planes $\bar{H}$ and $\bar{H}'$, respectively, then by (III.18) there is a σ in $U(\bar{H} \perp Y)$ with $\sigma x = x'$ and $\sigma y = y'$, i.e., $\sigma\bar{H} = \bar{H}'$. But σ will carry the orthogonal complement of $\bar{H}$ to the orthogonal complement of $\bar{H}'$, i.e., $\sigma Y = V$. Thus $V \simeq Y$.

Observe the above corollary does not require V to have hyperbolic rank ≥ 1 since σ is chosen from $U(\bar{H} \perp V)$ and the hyperbolic rank of $\bar{H} \perp V$ is ≥ 1.

(III.20) COROLLARY. (Generators of $U(V)$) The group $U(V)$ is generated by the isometries $\sigma_{u,x}$, $\sigma_{v,x}$, Δ and Φ_ε for various choices of x and ε.

Proof. The proof is similar in style to the previous proofs of this section and thus we only sketch it.

If $V = H$ is a hyperbolic plane, then it is straightforward to check that if σ is in $U(H)$ then there exists a product $\tau_1 \ldots \tau_t$ of the above maps with

$$\tau_1 \tau_2 \ldots \tau_t \sigma = i_H .$$

Thus, $\sigma = \tau_t^{-1} \ldots \tau_1^{-1}$.

The proof then proceeds by induction. If $\lambda = -1$ then $V = H_1 \perp \cdots \perp H_s$, H_i hyperbolic planes. Let σ be in $U(V)$ and suppose $H = \sigma H_s$. Then H is a hyperbolic plane and there is a product $\tau_1 \ldots \tau_t$ of the above isometries by (III.18) with $\tau_1 \ldots \tau_t \sigma = i_{H_s}$. But $\tau_1 \ldots \tau_t \sigma$: $H_s^{\perp} \to H_s^{\perp}$ where $H_s^{\perp} = H_1 \perp \cdots \perp H_{s-1}$. By induction, $\tau_1 \ldots \tau_t \sigma = \delta_1 \ldots \delta_r$, δ_r of the desired form in $U(H_s^{\perp})$. By defining δ_i to be the identity on H_s , we may extend δ_i to $U(V)$. Then

$$\sigma = \tau_t^{-1} \ldots \tau_1^{-1} \delta_1 \ldots \delta_r .$$

If $\lambda = 1$, then

$$V = H \perp Re_1 \perp \cdots \perp Re_s$$

where $\beta(e_i, e_i)$ is a unit. If σ is in $U(V)$ then $\beta(\sigma(e_s), \sigma(e_s)) = \beta(e_s, e_s)$. By (III.17) there is a product $\tau_1 \ldots \tau_t$ of desired generators such that $\tau_1 \ldots \tau_t \sigma e_s = e_s$. Again, $\tau_1 \ldots \tau_t \sigma : (Re_s^{\perp}) \to (Re_s^{\perp})$. Proceed by induction.

Let A be a proper ideal of R. The ring morphism $\Pi_A : R \to R/A$ induces a natural R-morphism $\Pi_A : V \to V/AV$. The R-morphism Π_A determines a group morphism

$$\lambda_A : U(V) \to U(V/AV)$$

given by

$$(\lambda_A\sigma)\Pi_A = \Pi_A\sigma ,$$

i.e., for x in V

$$(\lambda_A\sigma)(\Pi_A x) = \Pi_A(\sigma(x)) .$$

Observe V/AV is naturally a λ-inner product space under $\bar{\beta}$ where

$$\bar{\beta}(\Pi_A x,\Pi_A y) = \Pi_A(\beta(x,y)) .$$

Further, hyperbolic rank of $V/AV \geq$ hyperbolic rank of V.

<u>(III.21)</u> <u>COROLLARY</u>. If A is a proper ideal of R, the group morphism

$$\lambda_A : U(V) \to U(V/AV)$$

is surjective.

<u>Proof</u>. Since R/A is local the generators of $U(V/AV)$ are given in (III.20). The map $\Pi_A : R^* \to (R/A)^*$ on the units of R induced by $\Pi_A : R \to R/A$ is surjective. Thus, for each generator of $U(V/AV)$ one can find a generator of $U(V)$ mapping onto it. Hence, λ_A is surjective.

(III.22) THEOREM. Suppose $\dim(V) \geq 3$. Then the center of $U(V)$ is $\{\pm I\}$.

Proof. Suppose ϕ is in the center of $U(V)$. Then $\sigma\phi = \phi\sigma$ for all σ in $U(V)$.

We have $V = H \perp W$. Suppose x is in W and $\phi(x) = \alpha u + \delta v + y$ where α and δ are in R and y is in W . Then

$$\phi(x) = \phi(\Phi_\varepsilon x) .$$

Thus

$$\begin{aligned} \phi(x) &= \Phi_\varepsilon(\phi(x)) \\ &= \varepsilon\alpha u + \varepsilon^{-1}\delta v + y . \end{aligned}$$

Letting $\varepsilon = 2$, we have $\alpha - 2\alpha = 0$ and $\delta - \frac{1}{2}\delta = 0$. Hence $\alpha = \delta = 0$ and $\phi(x)$ is in W . Thus $\phi(W) \subseteq W$.

If x is unimodular in W (note $\dim(V) \geq 3$ hence $\dim(W) \geq 1$), select y in W with $\beta(x,y) = 1$. Consider $\sigma_{u,y}\phi = \phi\sigma_{u,y}$:

$$\begin{aligned} \phi(x) + \beta(y,\phi(x))u &= \sigma_{u,y}(\phi(x)) \\ &= \phi(\sigma_{u,y}(x)) \\ &= \phi(x + u) \\ &= \phi(x) + \phi(u) . \end{aligned}$$

Hence $\phi(u) = \eta u$ (η a unit). Similarly $\phi(v) = \bar{\eta}v$. Since $1 = \beta(u,v) = \beta(\phi(u),\phi(v)) = \eta\bar{\eta}\beta(u,v) = \eta\bar{\eta}$, we have $\bar{\eta} = \eta^{-1}$.

Let x be in W . Again $\sigma_{v,x}\phi = \phi\sigma_{v,x}$. Thus,

$$\begin{aligned}\sigma_{v,x}\phi(u) &= \sigma_{v,x}(\eta u)\\ &= \eta u - \eta x - \frac{1}{2}\lambda\eta\beta(x,x)v\end{aligned}$$

and

$$\begin{aligned}\phi\sigma_{v,x}(u) &= \phi[u - x - \frac{1}{2}\lambda\beta(x,x)v]\\ &= \eta u - \phi(x) - \frac{1}{2}\lambda\bar{\eta}\beta(x,x)v\end{aligned}$$

are equal. Hence, for all x in W , $\phi(x) = \eta x$. Select x and y in W with $\beta(x,y) = 1$. Then

$$1 = \beta(x,y) = \beta(\phi x,\phi y) = \eta^2\beta(x,y) = \eta^2$$

and $\eta = \pm 1$. This completes the proof.

Let A be an ideal of R with $A \neq R$. The morphism $\Pi_A : R \to R/A$ induces (III.21) a surjective group morphism $\lambda_A : U(V) \to U(V/AV)$. By (III.22) let

$$\begin{aligned}U(V,A) &= \lambda_A^{-1}\ (\text{Center } U(V/AV))\\ &= \lambda_A^{-1}\ \{\pm I\}\\ &= \{\sigma \text{ in } U(V) \mid \lambda_A\sigma = \pm I\}\end{aligned}$$

and call U(V,A) the <u>group of isometries of level</u> A <u>of</u> U(V) . For special cases: let $U(V,R) = U(V)$ and $U(V,0) = \{\pm I\}$.

Then $\sigma_{u,x}$ and $\sigma_{v,x}$ are in $U(V,O(x))$ where $O(x)$ is the order of x (see (II.A)). If ε is a unit of R with $\varepsilon \equiv 1$ modulo A , i.e., $\Pi_A \varepsilon = 1$, then Φ_ε is in $U(V,A)$.

The group generated by unitary transvections $\sigma_{u,x}$ and $\sigma_{v,y}$ is denoted by $EU(V)$ and called the _Eichler group_ (this is our choice of terminology — often this group is not given a name). The $EU(V)$-normal subgroup of $U(V)$ generated by isometries $\sigma_{u,x}$ and $\sigma_{v,x}$ with $O(x) \subset A$ (A an ideal of R) is denoted by $EU(V,A)$ and is called the _Eichler group of level_ A . Observe

$$EU(V,A) \leq U(V,A) .$$

Analogously with Chapter II, we would like to show the normal subgroups of $U(V)$ are trapped between $EU(V,A)$ and $U(V,A)$. However, due to technical problems we do not carry through the classification of normal subgroups in this context. We handle the symplectic and orthogonal groups separately. We conclude this section with a discussion of the transitivity and generators of $U(V,A)$.

(III.23) THEOREM. Let $\dim(V) \geq 3$. The group $U(V,A)$ acts as a transformation group on the set of unimodular vectors of V of the same norm. For a given norm the $U(V,A)$-orbits consist of those vectors x

and y with $x \equiv \pm y \bmod A$.

Proof. By (III.17) we may assume $A \neq R$. If ϕ is in $U(V,A)$ and $\phi x = y$ then $x \equiv \pm y \bmod A$.

Conversely, suppose $x \equiv \pm y \bmod A$. We want to carry x into y by a sequence of elements from $U(V,A)$. Since $-I$ is in $U(V,A)$, after an application of $-I$ (if necessary) we may assume $x \equiv y \bmod A$.

Recall $V = H \perp W$ where $H = Ru \oplus Rv$ is a hyperbolic plane. Let $x_1,\ldots,x_t$ be a basis for W . Select $y_1,\ldots,y_t$ in W (a "dual" basis) with $\beta(x_i,y_j) = \delta_{ij}$.

Suppose

$$x = \alpha u + \delta v + \sum a_i x_i$$

and

$$y = \bar{\alpha} u + \bar{\delta} v + \sum b_i x_i \ .$$

Since $x \equiv y \bmod A$, $\alpha \equiv \bar{\alpha}$, $\delta \equiv \bar{\delta}$ and $a_i \equiv b_i$ all modulo A .

The vector x is unimodular. Hence α , δ or some a_i is a unit. We show that we may assume α or δ is a unit.

If α or δ are non-units then (a) $\bar{\alpha}$ and $\bar{\delta}$ are non-units, (b) a_s is a unit for some s , $1 \leq s \leq t$, and thus b_s is a unit. Then

$$\sigma_{u,y_s} x \quad \text{and} \quad \sigma_{u,y_s} y$$

have units for their coefficients of u . If we manufacture a ϕ in $U(V,A)$ with

$$\phi\sigma_{u,y_s} x = \sigma_{u,y_s} y$$

then we are done since

$$\sigma_{u,y_s}^{-1} \phi\sigma_{u,y_s} x = y$$

and

$$\sigma_{u,y_s}^{-1} \phi\sigma_{u,y_s} \equiv \pm I \mod A .$$

Thus, suppose

$$x = \alpha u + \delta v + \sum a_i x_i ,$$
$$y = \bar{\alpha} u + \bar{\delta} v + \sum b_i x_i$$

where α is a unit (thus $\bar{\alpha}$ is a unit) and $a_i = b_i + q_i$ with q_i in A . A similar argument will apply if δ is a unit. Set $z = \sum \bar{\alpha}^{-1} q_i x_i$. Then $O(z) \subset A$ and hence $\sigma_{v,z}$ is in $U(V,A)$.

Thus

$$\sigma_{v,z} y = \bar{\alpha} u + \varepsilon v + \sum a_i x_i .$$

Apply Φ_α^{-1} to x and $\sigma_{v,z} y$,

(*)
$$\Phi_\alpha^{-1}x = u + \rho v + w ,$$
$$\Phi_\alpha^{-1}\sigma_{v,z}y = \mu u + \bar{\rho} v + w$$

where $w = \sum a_i x_i$. Then $\mu \equiv 1 \mod A$ and consequently Φ_μ^{-1} is in $U(V,A)$. Thus

(**)
$$\Phi_\mu^{-1}\Phi_\alpha^{-1}\sigma_{v,z}y = u + \hat{\rho} v + w .$$

Applying $\sigma_{v,w}$ to (*) and (**),

$$\sigma_{v,w}\Phi_\alpha^{-1}x = u + \zeta v$$
$$\sigma_{v,w}\Phi_\mu^{-1}\Phi_\alpha^{-1}\sigma_{v,z}y = u + \bar{\zeta} v .$$

If $\lambda = 1$, then

$$\beta(u + \zeta v, u + \zeta v) = \beta(u + \bar{\zeta} v, u + \bar{\zeta} v)$$

implies $\zeta = \bar{\zeta}$ and we are done. If $\lambda = -1$, a more complicated situation occurs. Let

$$X = u + \zeta v$$
$$Y = u + \bar{\zeta} v$$

and $\zeta = \bar{\zeta} - a$ with a in A . Define $\tau_a : H \to H$ by

$$\tau_a(z) = z + a\beta(v,z)v . \qquad \text{(Symplectic Transvection)}$$

Extend τ_a to V by $\tau_a|_W = 1$. Then $\tau_a \equiv I \mod A$ and

$\tau_a(Y) = (u + \bar{\zeta}v) + a\beta(u + \zeta v,v)v = u + \bar{\zeta}v + av = u + \zeta v = X$. Further, as in the discussion prior to (III.17),

$$\tau_a = \sigma_{v,(a/2)v} .$$

Thus, if $\phi = I$ when $\lambda = 1$ and $\phi = \tau_a$ when $\lambda = -1$, then

$$\Phi_\alpha \sigma^{-1}_{v,w} \phi \sigma_{v,w} \Phi^{-1}_\mu \Phi^{-1}_\alpha \sigma_{v,z} y = x$$

and we are done.

Let T denote the family of hyperbolic planes in V . If $H = Ru \oplus Rv$ is in T and ϕ is in $U(V,A)$ then $\phi H = \bar{H}$ where $\bar{H} = R\phi(u) \oplus R\phi(v)$. Clearly $\Pi_A H = \Pi_A \bar{H}$. We now show the converse of this fact.

<u>(III.24)</u> <u>THEOREM</u>. Let $\dim(V) \geq 3$. The group $U(V,A)$ acts as a transformation group on T the family of hyperbolic planes in V . Two planes H and $\bar{H}$ lie in the same $U(V,A)$-orbit if $\Pi_A H = \Pi_A \bar{H}$.

<u>Proof</u>. Let $H = Ru \oplus Rv$ and $\bar{H} = Re \oplus Rf$ with $\Pi_A H = \Pi_A \bar{H}$. Further, we may assume u , v , e and f are selected such that $\Pi_A u = \Pi_A e$ and $\Pi_A v = \Pi_A f$. Since $\beta(u,u) = \beta(e,e) = 0$ and $u \equiv e \mod A$, there is a σ in $U(V,A)$ with $\sigma u = e$. Thus, without loss we may assume $u = e$. Hence

$$H = Ru \oplus Rv$$

$$\bar{H} = Ru \oplus Rf$$

Then $f = \alpha u + \delta v + w$ where w is in W , $V = H \perp W$. Further, $\beta(u,f) = 1$ so $\delta = 1$ and

$$f = \alpha u + v + w .$$

The purpose ahead is to construct a σ in $U(V,A)$ with $\sigma u = u$ and $\sigma f = v$.

Since $f \equiv v \mod A$, we have α in A and $O(w) \subset A$. Therefore $\sigma_{u,\lambda w}$ is in $U(V,A)$ and

$$\sigma_{u,\lambda w}(u) = u$$
$$\sigma_{u,\lambda w}(f) = \delta u + v .$$

If $\lambda = 1$, $\beta(f,f) = 0$ implies

$$\begin{aligned} 0 &= \beta(\sigma_{u,\lambda w}f,\sigma_{u,\lambda w}f) \\ &= \beta(\delta u + v,\delta u + v) \\ &= 2\delta \end{aligned}$$

and $\delta = 0$. Thus if $\lambda = 1$ we are done.

Suppose $\lambda = -1$. Analogous to the final step in the proof of (III.23) take τ_δ where

$$\tau_\delta(z) = z + \delta\beta(z,u)u .$$

Then $\tau_\delta(u) - u$ and $\tau_\delta(\delta u + v) = v$. Note $f \equiv v \mod A$ implies δ is in A . This finishes the proof.

(III.25) THEOREM. Let A ($\neq R$) be an ideal of R and let $\dim(V) \geq 3$. Then $U(V,A)$ is generated by $-I$, $EU(V,A)$ and Φ_ε where $\varepsilon \equiv 1$ modulo A.

Proof. We treat two cases $\lambda = 1$ and $\lambda = -1$.

Suppose $\lambda = -1$. The proof is by induction on $\dim(V)$. Since $\dim(V) \geq 3$, $V = H \perp W$ where $\dim(W) \geq 1$, in fact $\dim(W) = 2t$ for some t (by (III.13)) and W is an orthogonal sum of hyperbolic planes: $W = \bar{H}_1 \perp \cdots \perp \bar{H}_t$. Let σ be in $EU(V,A)$. Then $\hat{H} = \sigma\bar{H}_t$ is a hyperbolic plane with $\hat{H} \equiv \bar{H}_t$ modulo A. By (III.24) there is a ϕ (observe the key idea is that, using (III.16), (III.23) and (III.24) were proven by employing only products of isometries of the above form) which is a product of the above isometries with $\phi\hat{H} = \bar{H}_t$, i.e., $\phi\sigma\bar{H}_t = \bar{H}_t$. Then $\phi\sigma$ carries the orthogonal complement of $\bar{H}_t$ to the orthogonal complement of $\bar{H}_t$. Thus, by induction we may assume we have constructed a product ϕ of isometries of the above form satisfying $\phi\sigma|_W =$ identity. Employing the techniques of (III.23) and (III.24), we may construct additional products of isometries, say τ, with $\tau\phi\sigma =$ identity on V. Hence

$$\sigma = \phi^{-1}\tau^{-1} .$$

Suppose $\lambda = 1$. Again the proof is induction on $\dim(V)$. Since $\lambda = 1$, by (III.7)

$$V = H \perp W$$
$$= H \perp \langle u_1 \rangle \perp \cdots \perp \langle u_t \rangle$$

where $\{x_1,\ldots,x_t\}$ is a basis of W and $\beta(x_i,x_i) = u_i$. Let σ be in $U(V,A)$. Then σx_t is unimodular and $\sigma x_t \equiv \pm x_t$ modulo A. By the proof of (III.23) there is a product ϕ of isometries of the above form with $\phi\sigma x_t = x_t$. Thus $\phi\sigma : (Rx_t^{\perp}) \to (Rx_t^{\perp})$. Again by induction we reduce the problem to the action of σ on H and likewise isometries of the desired form may be constructed reducing σ to the identity. This gives the theorem.

(E) HYPERBOLIC SPACES: SPLIT SYMMETRIC AND SYMPLECTIC

In this section we assume 2 is a unit of R. Hence $\lambda = \pm 1$.

Further, we assume (V,β) is a split or hyperbolic space (see Section (III.C)). Thus, a basis of V may be selected such that relative to this basis the matrix of β has the form

$$\begin{bmatrix} 0 & \lambda I \\ I & 0 \end{bmatrix}.$$

Equivalently,

$$(V,\beta) \simeq H_1 \perp H_2 \perp \cdots \perp H_m$$

where the H_i are hyperbolic planes and $\beta|_{H_i \times H_i}$ has matrix

$$\begin{bmatrix} 0 & \lambda \\ 1 & 0 \end{bmatrix} .$$

If $\lambda = -1$ the space (V,β) is called _symplectic_ and the group $U(V)$ is called the _symplectic group_ and denoted $Sp(V)$. If $\lambda = 1$, (V,β) has not been supplied a proper name. (Recently Snapper [26] suggested that these spaces be named Artinian spaces. However, this choice has met with some argument.) For this discussion, we shall call this space _split symmetric_ and the group $U(V)$, called the _split symmetric group_, will be denoted by $SS(V)$. Bak [35] has shown that for many purposes $Sp(V)$ and $SS(V)$ should be treated simultaneously (however, occasionally $Sp(V)$ will give sharper results).

This section is devoted to describing some of the elements of $Sp(V)$ or $SS(V)$. For convenience, we summarize the notation:

> 2 is a unit of R,
>
> $\lambda = \pm 1$,
>
> (V,β) is a split space
>
> Σ denotes either $Sp(V)$ (if $\lambda = -1$) or $SS(V)$ (if $\lambda = 1$).

Further, we assume a basis $\bar{B}$ of V has been selected such that the matrix of β relative to $\bar{B}$ is

$$\begin{bmatrix} 0 & \lambda I \\ I & 0 \end{bmatrix} .$$

(III.26) THEOREM. Let $\sigma = \begin{bmatrix} A & B \\ C & D \end{bmatrix}$ be in $GL(V)$. Then σ is in $\sum$ if and only if

(a) $A(D^t) + \lambda B(C^t) = I$,

(b) $B(A^t) = -\lambda A(B^t)$, and

(c) $D(C^t) = -\lambda C(D^t)$.

Proof. If $\begin{bmatrix} A & B \\ C & D \end{bmatrix}$ is in $\sum$ then

$$\begin{bmatrix} 0 & \lambda I \\ I & 0 \end{bmatrix} = \begin{bmatrix} A & B \\ C & D \end{bmatrix}\begin{bmatrix} 0 & \lambda I \\ I & 0 \end{bmatrix}\begin{bmatrix} A & B \\ C & D \end{bmatrix}^t$$
$$= \begin{bmatrix} BA^t + \lambda AB^t & BC^t + \lambda AD^t \\ DA^t + \lambda CB^t & DC^t + \lambda CD^t \end{bmatrix} .$$

Equating block matrices, the result follows.

Equations (b) and (c) in the above theorem lead to the following definition.

Definition. If A is an m by m matrix over R then $A^0 = -\lambda A^t$. Then A is 0-symmetric if $A^0 = A$.

Hence (b) and (c) may be stated as

(b') BA^t is 0-symmetric,

(c') DC^t is 0-symmetric.

To describe $\sigma = \begin{bmatrix} A & B \\ C & D \end{bmatrix}$ more carefully, recall the matrix for the form β, i.e., $\begin{bmatrix} 0 & \lambda I \\ I & 0 \end{bmatrix}$, arose from a splitting of V as $V = W \oplus U$ where $W = W^{\perp}$. Further U may be identified with W^*. Thus

$$V = W \oplus W^* .$$

The matrix $\begin{bmatrix} 0 & \lambda I \\ I & 0 \end{bmatrix}$ arises by selecting a basis $\{b_1,\ldots,b_m\}$ for W and $\{b_1^*,\ldots,b_m^*\}$ $(n = 2m)$ for W^* satisfying

$$\begin{aligned} \beta(b_i,b_j) &= 0 , \\ \beta(b_i^*,b_j^*) &= 0 , \\ \beta(b_i,b_j^*) &= b_j^*(b_i) = \lambda\delta_{ij} , \\ \beta(b_j^*,b_i) &= \delta_{ij} \end{aligned}$$

for $1 \le i,j \le m$. The isometry $\sigma : W \oplus W^* \to W \oplus W^*$,

$$\sigma = \begin{bmatrix} A & B \\ C & D \end{bmatrix} ,$$

may now be thought of as composed of linear maps

$$\begin{aligned} A &: W \to W , \\ B &: W^* \to W , \\ C &: W \to W^* , \\ \text{and} \quad D &: W^* \to W^* \end{aligned}$$

satisfying the conditions of the above theorem.

Let

$$(W,W^*)^0 = \{\alpha : W \to W^* \mid \alpha^0 = \alpha\}$$
$$= \{\alpha \text{ in } \mathrm{Hom}_R(W,W^*) \mid \alpha \text{ is 0-symmetric}\}$$

$$(W^*,W)^0 = \{\alpha : W^* \to W \mid \alpha^0 = \alpha\}$$
$$= \{\alpha \text{ in } \mathrm{Hom}_R(W^*,W) \mid \alpha \text{ is 0-symmetric}\} .$$

Then $(W,W^*)^0$ and $(W^*,W)^0$ are additive groups.

Define

$$\underline{\chi} : (W,W^*)^0 \to \Sigma$$

by

$$\underline{\chi} : \alpha \to \begin{bmatrix} I & 0 \\ \alpha & I \end{bmatrix}$$

and

$$\bar{\chi} : (W^*,W)^0 \to \Sigma$$

by

$$\bar{\chi} : \alpha \to \begin{bmatrix} I & \alpha \\ 0 & I \end{bmatrix} .$$

It is easy to see $\underline{\chi}$ and $\bar{\chi}$ are group morphisms from additive groups to mulitplicative subgroups of Σ . The fact that $\underline{\chi}(\alpha)$ and $\bar{\chi}(\alpha)$ are in Σ follows from (III.26). Thus,

$$\underline{\chi}(\alpha + \delta) = \underline{\chi}(\alpha)\underline{\chi}(\delta)$$

and

$$\underline{\chi}(-\alpha) = \underline{\chi}(\alpha)^{-1}$$

and similarly for $\bar{\chi}$.

We now generalize the isometry Φ_ε of Section (D).

Suppose $\varepsilon : W \to W$ is invertible. Then (see (I.F), for example) ε induces an invertible morphism $\varepsilon^* : W^* \to W^*$ called the <u>transpose</u> of ε . Indeed, selecting a basis for W with dual basis for W^* , then the matrix $[\varepsilon^*]$ of ε^* is precisely the transpose $[\varepsilon]^t$ of the matrix $[\varepsilon]$ of ε . The inverse of ε^* , $(\varepsilon^*)^{-1}$, is called the <u>contragredient</u> of ε and denoted $\check{\varepsilon}$. (See (II.F).)

Define

$$\Phi : GL(W) \to \sum$$

by

$$\Phi : \varepsilon \to \Phi_\varepsilon = \begin{bmatrix} \varepsilon & 0 \\ 0 & \check{\varepsilon} \end{bmatrix} .$$

A direct calculation shows Φ_ε is in $\sum$.

<u>(III.27)</u> <u>LEMMA</u>. (a) $\Phi_\varepsilon^{-1} = \Phi_{\varepsilon^{-1}}$.

(b) $\Phi_\varepsilon \bar{\chi}(\alpha)\Phi_\varepsilon^{-1} = \bar{\chi}(\varepsilon\alpha\varepsilon^*)$.

(c) $\Phi_\varepsilon \underline{\chi}(\alpha)\Phi_\varepsilon^{-1} = \underline{\chi}(\check{\varepsilon}\alpha\varepsilon^{-1})$.

Hence, the image of Φ normalizes the images of $\bar{\chi}$ and $\underline{\chi}$.

The remaining class of elements of $\sum$ which we examine are constructed from R-isomorphisms $\phi : W^* \to W$. Similar to the above, ϕ induces an isomorphism $\phi^* : W \to W^*$. Again, denote $(\phi^*)^{-1}$ by $\check{\phi}$. Define

$$w : \{\phi : W^* \to W \mid \phi \text{ isomorphism}\} \to \sum$$

by

$$w : \phi \to w(\phi) = \begin{bmatrix} 0 & \phi \\ \lambda\check{\phi} & 0 \end{bmatrix} .$$

Since

$$\lambda\phi(\lambda\check{\phi})^* = I ,$$

$w(\phi)$ is in $\sum$ by (III.27).

<u>(III.28)</u> <u>LEMMA.</u>

(a) $w(\phi)^2 = \Phi_{\lambda\phi\check{\phi}}$.

(b) $w(\phi)^{-1} = w(\lambda\phi^*)$.

(c) $w(\phi)\Phi_\varepsilon w(\phi)^{-1} = \Phi_{\phi\check{\varepsilon}\phi^{-1}}$.

(d) $w(\phi)\bar{X}(\alpha)w(\phi)^{-1} = \underline{X}(\lambda\check{\phi}\alpha\phi^{-1})$.

(e) $w(\phi)\underline{X}(\alpha)w(\phi)^{-1} = \bar{X}(\lambda\phi\alpha\phi^*)$.

Additional detail on the mappings $\bar{X}(\alpha)$, $\underline{X}(\alpha)$, Φ_ε and $w(\phi)$ is given in [8]. These isometries allow a description of the stable behavior of $\sum$ as $n \to \infty$. (See [8].)

(F) DISCUSSION

The early presentation and terminology of this section is based primarily on the discussions in Chapter I of [20] by Milnor and in [8] by Bass. We did not choose to introduce rings with involution and Hermitian forms. For a unified treatment of this together with modifications to handle the difficult case of characteristic 2 , i.e., 2 not a unit, one should see Bass [8]. Bass' paper provides an exposition of the foundations of the theory of non-singular Hermitian forms over rings with involution. A treatment of Hermitian forms and the unitary group over local rings is given by Baeza in [30]. The basic theory of projective modules which carry Hermitian forms over semi-local rings is provided in [94]. There are a number of sources providing an introduction to symmetric bilinear forms over rings, e.g., [8], [19], [22], and over fields, e.g., [1], [18], [26]. A matrix approach over the integers to these topics is given in [21].

Hyperbolic planes and spaces have proven to be crucial in the classification problem of inner product spaces. It has been said [26] that Artin first recognized the importance of these spaces. To assume that the inner product space V (say $\lambda = 1$) splits as $V = H \perp W$ where H is a hyperbolic place, induces a 'linearization' on the associated quadratic form. In a sense, allows linear techniques to be employed.

The "transvections" introduced in (D) have also been called Siegel-Transformations[1]

---

[1] See earlier footnote.

as they were perhaps first introduced in matrix form by Siegel.[1] Later Eichler [16] reformulated these transvections and illustrated their importance. Clearly a difficult problem is the description of the generators of the orthogonal group $O(V)$ of a space V over a ring. Without such a description it is difficult to determine the action of $O(V)$ on elements of V as a transformation group. Observe (III.18) through (III.21) depend on the transitivity of $O(V)$ on vectors of the same norm. An evolved form for these "transvections" and the above related questions is given in [8] and [35].

Notice the symplectic case is better behaved than the orthogonal. From the standpoint of bilinear forms it is not immediately clear why this should be. However, in the context of quadratic forms and their theory, the symplectic form induces a trivial quadratic form and consequently, the heart of the theory of quadratic forms arises from the study of symmetric bilinear forms. Indeed, if 2 is a unit these two theories are indistinguishable.

It should be noted that symmetric and symplectic forms arise naturally in the classification of bilinear forms when one imposes an orthogonality constraint, i.e., that $\beta(x,y) = 0$ implies $\beta(y,x) = 0$. For example, see Artin [1] pp. 110-113 or Biggs [11] pp. 42-47.

---

[1] These transformations have also been attributed to Dickson [13].

IV. THE SYMPLECTIC GROUP: NORMAL SUBGROUPS AND AUTOMORPHISMS

(A) INTRODUCTION

The purpose of this chapter is to provide a description of the normal subgroups and automorphisms of the symplectic group. We shift from the generators of $U(V)$ described in the last chapter to a more efficient generating set — the symplectic transvections. These were mentioned in the last chapter when we needed to move vectors (for $\lambda = -1$) around in hyperbolic planes.

To obtain a description of the normal subgroups we employ the induction technique of Klingenberg [91]. Thus, we need to understand the normal subgroups of the symplectic group of a hyperbolic plane. We show this may be reduced to a classification of the normal subgroups of $SL(V)$ when $\dim(V) = 2$. The normal subgroups of $SL(V)$ for $\dim(V) = 2$ are determined in the Appendix to this chapter subject to the assumptions R/m is not the field of 3 elements and 2 is a unit of R.

The automorphism theory is approached by way of [103] and is based on the transvection techniques of O'Meara. Here lines in the space will be identified with families of transvections. Then the families of transvections will be carried via the automorphism to families of transvections. This will establish a projectivity and the Fundamental Theorem of

Projective Geometry will be invoked to create the automorphism. Since a portion of this utilizes the normal subgroup theory the same assumptions are necessary, i.e., 2 is a unit of R and R/m is not the field of 3 elements. Further, we will assume $\dim(V) \geq 6$.

Throughout this chapter, R denotes a local ring having 2 a unit, maximal ideal m and residue class field $k = R/m$.

Let V be an R-space of dimension n and $\beta : V \times V \to R$ a symplectic inner product. (Hence we have the results of (III) with $\lambda = -1$ — since this context is fixed, the symbol λ as it appears in this chapter will be used to denote an element of R .)

If e and f are unimodular in V and satisfy $\beta(e,f) = 1$, then the plane $H = Re \oplus Rf$ is a symplectic inner product space and is called a <u>hyperbolic plane</u> (see (III.C)). The pair $\{e,f\}$ is called <u>hyperbolic</u>. Any unimodular element e in V may be complemented in V to a hyperbolic pair and V is an orthogonal direct sum

$$V = H_1 \perp \cdots \perp H_m$$

of hyperbolic planes. Hence $\dim(V) = n = 2m$. Further, two symplectic spaces of the same dimension are isometric.

Suppose $B = \{e_1,f_1,e_2,f_2,\ldots,e_m,f_m\}$ is a basis of V satisfying

$\beta(e_i,f_i) = 1$, $1 \le i \le m$, (thus $\beta(f_i,e_i) = -1$) and all other combinations of basis elements yielding 0 . Then the matrix of the form β is given by

$$\mathrm{Mat}(\beta) = \begin{bmatrix} 0 & 1 & & & & & & 0 \\ -1 & 0 & & & & & & \\ & & 0 & 1 & & & & \\ & & -1 & 0 & & & & \\ & & & & \ddots & & & \\ & & & & & & 0 & 1 \\ & 0 & & & & & -1 & 0 \end{bmatrix}$$

and the basis B is called a hyperbolic basis. If the above basis is listed as

$$\{f_1,f_2,\ldots,f_m,e_1,\ldots,e_m\}$$

then the matrix of β is

$$\mathrm{Mat}(\beta) = \begin{bmatrix} 0 & -I_m \\ I_m & 0 \end{bmatrix}$$

(see (III.E)) and the basis is called symplectic.

An R-morphism $\sigma : V \to V$ is called an isometry if

(a) σ is in $GL(V)$

and

(b) $\beta(\sigma x, \sigma y) = \beta(x,y)$ for all x and y in V.

Let $Sp(V)$ denote the group of isometries of (V,β).

Let A be an ideal of R. The ring morphism $\Pi_A : R \to R/A$ induces an R-morphism $\Pi_A : V \to V/AV$ of symplectic spaces. If $V = (V,\beta)$ then V/AV is a symplectic space $(V/AV, \bar{\beta})$ where

$$\bar{\beta}(\Pi_A x, \Pi_A y) = \Pi_A \beta(x,y) .$$

Then Π_A induces a group morphism

$$\lambda_A : Sp(V) \to Sp(V/AV)$$

by

$$(\lambda_A \sigma)(\Pi_A x) = \Pi_A(\sigma x) .$$

By (III.21), $\lambda_A : Sp(V) \to Sp(V/AV)$ is surjective.

(B) SYMPLECTIC TRANSVECTIONS

Let R be a local ring in which 2 is a unit. Let (V,β) be a symplectic space over R of dimension $\dim(V) = n = 2m$.

In this section we introduce isometries which prove to be a more convenient generating set than those described in (III).

(IV.1) THEOREM. (Characterization of Symplectic Transvections) Let τ be in $Sp(V)$. Then the following are equivalent:

(a) $\tau x = x + \lambda\beta(a,x)a$ for all x in V where λ is a fixed element of R and a is a fixed unimodular vector in V.

(b) There is a line L in V satisfying $\tau x - x$ is in L for all x in V.

(c) There is a hyperplane H in V with $\tau|_H$ equal to the identity on H.

An element τ satisfying any of the above equivalent conditions is called a symplectic transvection and denoted $\tau = \tau_{a,\lambda}$.[1]

Proof. We first show (b) implies (c). Let $L = Ra$ be such that $\tau x - x$ is in L for all x in V. Let $H = L^{\perp}$. By (III.8), H is a hyperplane.

Let y be in H. For any x in V,

$$0 = \beta(\tau x - x,y) = \beta(\tau x,y) - \beta(x,y) .$$

Thus, $\beta(\tau x,y) = \beta(x,y)$. Then, since τ, and hence τ^{-1}, is in $Sp(V)$,

$$\beta(x,\tau^{-1}y) = \beta(x,y)$$

---

[1] The symbol λ denotes an element of R rather than -1 as in the last chapter. See the remark in the Introduction.

for all x in V. Thus $\tau^{-1}y = y$. Equivalently, $\tau y = y$ for all y in H and $\tau|_H$ = identity.

Reversing the above argument shows (c) implies (b). We show (c) implies (a). Let τ be in $Sp(V)$ and suppose $\tau|_H$ = identity on a hyperplane H. Let $L = Ra = H^\perp$. Since a, a generator of a line, is unimodular, a may be complemented by b satisfying $Ra \oplus Rb$ is a hyperbolic plane (see (III.14)). Let

$$V = (Ra \oplus Rb) \perp P .$$

By the previous argument, $\tau b - b$ is in $L = Ra$. Thus $\tau b = b + \lambda a$ for some λ in R. We now claim that for all x in V

$$\tau x = x + \lambda\beta(a,x)a .$$

If x is in V then $x = ra + sb + p$ for r and s in R and p in P. Then

$$\begin{aligned} \tau x &= r\tau a + s\tau b + \tau p \\ &= ra + s\tau b + p \qquad \text{(since } a \text{ and } p \text{ are in} \\ &= ra + sb + s\lambda a + p \\ &= (ra + sb + p) + \lambda\beta(a,ra + sb + p)a \\ &= x + \lambda\beta(a,x)a \quad . \end{aligned}$$

To show (a) implies (c) is straightforward.

Let $\tau = \tau_{a,\lambda}$ be a symplectic transvection. The line $L = Ra$ is called the <u>line of</u> τ and the hyperplane $H = (Ra)^{\perp}$ is called the <u>hyperplane of</u> τ. If λ is a unit in R, then $\tau = \tau_{a,\lambda}$ is called a <u>unimodular</u> symplectic transvection. When $\tau = \tau_{a,\lambda}$ is unimodular then $L = Ra$ and $H = (Ra)^{\perp} = \{x \text{ in } V \mid \beta(x,a) = 0\} = \{x \text{ in } V \mid \beta(x,L) = 0\}$ are uniquely determined.

$\tau_{a,\lambda}$ unimodular:

$L = Ra$

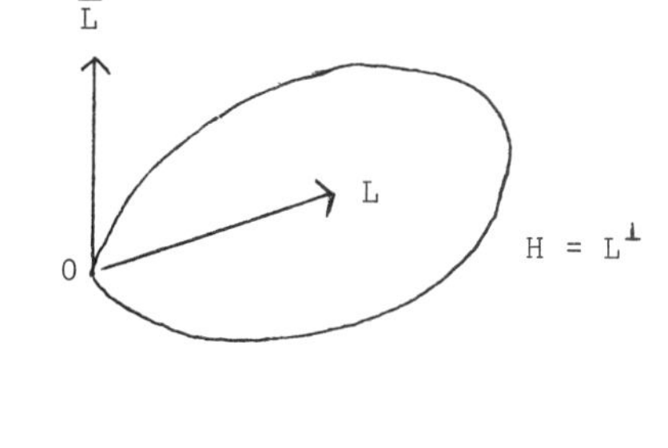

$$V = \bar{L} \oplus H$$
$$= \underbrace{(L \oplus \bar{L})}_{\text{hyperbolic plane}} \perp P$$

In (III.D) we discussed a collection of symmetries $\sigma_{u,x} : V \to V$ called unitary transvections. In fact, symplectic transvections actually occurred in several proofs. We now describe precisely the relationship between $\tau_{a,\lambda}$ and $\sigma_{u,x}$.

As in (III), let u be unimodular and x be in V with $\beta(u,x) = 0$. For z in V, recall

$$\sigma_{u,x}(z) = z + \beta(u,z)x + \beta(x,z)u .$$

We construct a product τ of symplectic transvections satisfying

$\sigma_{u,x} = \tau$. The construction of τ is somewhat messy.

First, let $\{u = u_1, v_1\} \cup \{u_2, v_2\} \cup \cdots \cup \{u_m, v_m\}$ be a hyperbolic basis of V . We construct τ so that

$$\sigma_{u,x}(u_i) = \tau(u_i)$$
$$\sigma_{u,x}(v_i) = \tau(v_i)$$

for $1 \leq i \leq m$. Hence $\sigma_{u,x} = \tau$. To do this observe, if

$$x = (\delta_1 u_1 + 0v_1) + (\delta_2 u_2 + \alpha_2 v_2) + \cdots + (\delta_m u_m + \alpha_m v_m) .$$

(Note $\beta(u,x) = 0$ implies coefficient of v_1 in x is 0), then

$$\sigma_{u,x}(u_1) = u_1$$
$$\sigma_{u,x}(v_1) = v_1 + x + \delta_1 u_1$$

and, for $i \geq 2$,

$$\sigma_{u,x}(u_i) = u_i - \alpha_i u_1$$
$$\sigma_{u,x}(v_i) = v_i + \delta_i u_1 .$$

First consider $i \geq 2$:

$$\tau_{u_i,-\delta_i}\tau_{u_1+u_i,\delta_i} \quad \begin{cases} u_1 \to u_1 \\ v_1 \to v_1 + \delta_i u_1 + \delta_i u_i \\ \\ u_i \to u_i \\ v_i \to v_i + \delta_i u_1 \\ \\ u_j \to u_j & j \neq i,1 \\ v_j \to v_j \end{cases}$$

and, in turn,

$$\tau_{v_i,-\alpha_i}\tau_{u_1+v_i,\alpha_i} \quad \begin{cases} u_1 \to u_1 \\ (v_1 + \delta_i u_1 + \delta_i u_i) \\ \quad \to v_1 + [\delta_i+\alpha_i(1-\delta_i)]u_1 + \delta_i u_i + \alpha_i v_i \\ \\ v_i + \delta_i u_1 \to v_i + \delta_i u_1 \\ u_i \to u_i - \alpha_i u_1 \\ \\ u_j \to u_j & j \neq i,1 \\ v_j \to v_j \end{cases}$$

Thus, set

$$\tau_1(i) = \tau_{u_i,-\delta_i}\tau_{u_1+u_i,\delta_i}$$

$$\tau_2(i) = \tau_{v_i,-\alpha_i}\tau_{u_1+v_i,\alpha_i}$$

and

$$\tau_i = \tau_2(i)\tau_1(i)$$

for $2 \le i \le m$.

The "desired" τ is given by

$$\tau = \prod_{i=2}^{m} \tau_i \ .$$

A direct computation shows τ agrees with $\sigma_{u,x}$ on each basis vector except possibly v_1 . Here

$$\begin{aligned} \tau(v_1) &= v_1 + [\textstyle\sum_{i=2}^{m} (\delta_i + \alpha_i(1 - \delta_i)]u_1 + \sum_{i=2}^{m} \delta_i u_i + \sum_{i=2}^{m} \alpha_i v_i \\ &= v_1 + (\text{——})u_1 + x - \delta_1 u_1 \\ &= v_1 + \lambda u_1 + x \end{aligned}$$

where $\lambda = \sum (\delta_i + \alpha_i(1 - \delta_i)) - \delta_1$ while

$$\sigma_{u,x}(v_1) = v_1 + \delta_1 u_1 + x \ .$$

But this difference occurs only in coordinate u_1 . Thus, multiply τ by

$$\tau_{u_1,\overline{\lambda}}$$

where $\overline{\lambda} = 2\delta_1 - \sum (\delta_i + \alpha_i(1 - \delta_i))$. The action of $\tau_{u_1,\overline{\lambda}}$ fixes the values of the other coordinates and

$$\tau_{u_1,\overline{\lambda}}\tau(v_1) = v_1 + \delta_1 u_1 + x \ .$$

Thus,

$$\sigma_{u,x} = \tau_{u_1,\bar{\lambda}}\tau$$

for τ given above. A similar formula will give $\sigma_{v,x}$ as a product of transvections.

If A is the ideal generated by the coefficients of x in $\sigma_{u,x}$ then

$$\lambda_A : \sigma_{u,x} \to I$$

under $\lambda_A : Sp(V) \to Sp(V/AV)$. That is, the order of $\sigma_{u,x}$, $O(\sigma_{u,x}) = A$ (in the sense of (II.A) or look ahead to the next section). But

$$x = \delta_1 u_1 + (\delta_2 u_2 + \alpha_2 v_2) + \cdots + (\delta_m u_m + \alpha_m v_m)$$

and thus $\delta_1, \delta_2, \ldots, \delta_m$ and $\alpha_2, \ldots, \alpha_m$ are in A. Thus, each transvection appearing in the above product $\tau_{u_1,\bar{\lambda}}\tau$ will reduce to the identity modulo A. Hence their orders are contained in A. We will utilize this observation in the next section.

We examine the remaining isometries Δ and Φ_ε of (III). Their action was confined to a hyperbolic plane $H = Ru \oplus Rv$ splitting V where $\beta(u,v) = 1$.

The isometry $\Phi_\varepsilon : H \to H$ was given by $\Phi_\varepsilon(u) = \varepsilon u$ and $\Phi_\varepsilon(v) = \varepsilon^{-1} v$ for ε a unit of R. Consider

$$\tau_1 = \tau_{u+v,1-\varepsilon} ,$$

$$\tau_2 = \tau_{v,\varepsilon^{-1}(\varepsilon-1)} \quad \text{and}$$

$\tau_3 = \tau_{u,\varepsilon(\varepsilon-1)}$. Then

$$\tau_3\tau_2\tau_1(u) = \varepsilon u$$
$$\tau_3\tau_2\tau_1(v) = \varepsilon^{-1}v \ .$$

Observe if $\varepsilon \equiv 1$ modulo A then $\tau_1 \equiv \tau_2 \equiv \tau_3 \equiv I$ (modulo A) .

The isometry $\Delta : H \to H$ was given by $\Delta(u) = v$ and $\Delta(v) = -u$. A direct computation gives

$$\Delta = \tau_{v,-2}\tau_{u-v,-1} \ .$$

<u>(IV.2)</u> <u>PROPOSITION</u>. The isometries $\sigma_{u,x}$, $\sigma_{v,x}$, Φ_ε and Δ given in (III) may be written as products of symplectic transvections. Further, if A is an ideal of R such that $\sigma_{u,x}$, $\sigma_{v,x}$ or Φ_ε is congruent to I modulo A , i.e., carried to I under λ_A , then the symplectic transvections are congruent to I modulo A .

We pause to observe a general technique appearing in the above proofs. Suppose u , v and w are vectors with $\beta(u,v) = 1$ and $\beta(u,w) = \beta(v,w) = 0$.

Then
$$\tau_{u,-\alpha}\tau_{w+u,\alpha}(v) = v + \alpha w$$
$$\tau_{v,-\alpha}\tau_{w+v,\alpha}(u) = u - \alpha w \ ,$$

i.e., to alter u and v , the first transvection adds too much then the seco subtracts off a little giving a net change of $\pm\alpha w$ (α arbitrary in R).

The isometries $\sigma_{u,x}$, $\sigma_{v,x}$, Φ_ε and Δ determined (III.17), (III.18) and (III.20). Since these isometries are products of transvections, the results may be restated in the following theorem.

(IV.3) THEOREM.

(a) The symplectic group $Sp(V)$ is generated by symplectic transvections. (III.20)

(b) The group $Sp(V)$ is transitive on unimodular vectors in V . Equivalently, $Sp(V)$ is transitive on lines in V . (III.17)

(c) The group $Sp(V)$ is transitive on hyperbolic planes in V . (III.18)

Theorem (III.22) determined the center of $Sp(V)$ provided $\dim(V) \geq 3$. We now remove this restriction, i.e., we show the center of $Sp(V)$ is $\pm I$ for $\dim(V) \geq 2$.

(IV.4) PROPOSITION. The center of $Sp(V)$ is $\{I,-I\}$.

Proof. Let ρ be in the center of $Sp(V)$ and $\dim(V) = 2$. We show $\rho x = x$ or $\rho x = -x$ for all x in V . Let a be unimodular in V and $\tau = \tau_{a,1}$ be a symplectic transvection. Then $\rho\tau\rho^{-1} = \tau$ since ρ is in the center, but $\rho\tau\rho^{-1} = \tau_{\rho a,1}$ by direct computation.

Since $\tau_{a,1} = \tau_{\rho a,1}$, $\rho a = \pm a$. Select b in V such that b is unimodular and $\{a,b\}$ is R-free. Then

$$\rho b = \pm b$$

$$\rho(a + b) = \pm(a + b)$$

and

$$\rho(a + b) = \rho a + \rho b .$$

By comparing results one concludes the sign of ρa must agree with the sign of ρb. Hence $\rho x = \pm x$ for all x in a basis of V and, consequently, $\rho x = \pm x$ for all x in V, i.e., $\rho = \pm I$.

We have adopted a "transvectional" approach to the structure theory and automorphism theory of $Sp(V)$. The next result summarizes some useful observations on symplectic transvections.

(IV.5) THEOREM. (On Symplectic Transvections)

(a) $\tau_{a,\lambda}^{-1} = \tau_{a,-\lambda}$.

(b) $\tau_{a,\lambda}\tau_{b,\mu}(x) = x + (\lambda\beta(a,x)a + \mu\beta(b,x)b) + \lambda\mu\beta(a,b)\beta(b,x)a$.

In particular, $\tau_{a,\lambda}\tau_{a,\mu} = \tau_{a,\lambda+\mu}$.

(c) $\sigma\tau_{a,\lambda}\sigma^{-1} = \tau_{\sigma a,\lambda}$ for σ in $Sp(V)$.

(d) If $\tau_{a,\lambda}$ is a unimodular symplectic transvection then $\tau_{a,\lambda} = \tau_{b,\mu}$ implies there is a unit α with

$$a = \alpha b \quad \text{and} \quad \alpha^{-2}\mu = \lambda .$$

(e) Let τ_1 and τ_2 be unimodular symplectic transvections. Then τ_1 and τ_2 commute if and only if their lines are orthogonal.

(f) Let $\tau = \tau_{a,\lambda}$ be a unimodular symplectic transvection and let $C(\tau) = \{\sigma \text{ in } Sp(V) \mid \sigma\tau = \tau\sigma\}$ denote the centralizer of τ in $Sp(V)$. Then

$$C(\tau) = \{\sigma \text{ in } Sp(V) \mid \sigma a = \pm a\} .$$

(g) If τ_1 and τ_2 are symplectic transvections with the same line then $\tau_1\tau_2$ is a symplectic transvection.

(h) Let τ_1 and τ_2 be unimodular symplectic transvections. Then $C(\tau_1) = C(\tau_2)$ if and only if τ_1 and τ_2 have the same line.

Proof. Statements (a), (b) and (c) are immediate.

To show (d), suppose $\tau_{a,\lambda}(x) = \tau_{b,\mu}(x)$ for all x in V. Thus $\lambda\beta(a,x)a = \mu\beta(b,x)b$ for all x. Since a is unimodular there is a y with $\beta(a,y) = 1$. Thus $a = \lambda^{-1}\mu\beta(b,y)b$. Similarly, there is a z with $\beta(b,z) = 1$. Hence $a = \lambda^{-1}\mu\beta(b,y)b = \lambda^{-1}\lambda\beta(a,z)\beta(b,y)a$ which implies $\beta(a,z)\beta(b,y) = 1$. Let $\alpha = \mu\lambda^{-1}\beta(b,y)$. Then $a = \alpha b$. Further $\beta(b,y)^{-1} = \beta(a,z) = \beta(\alpha b,z) = \alpha\beta(b,z) = \alpha$. Thus α and hence μ is a unit. Then

$$\begin{aligned}\alpha^{-2}\mu &= (\mu\lambda^{-1}\beta(b,y))^{-2}\mu \\ &= \beta(b,y)^{-2}\lambda^{2}\mu^{-2}\mu \\ &= \beta(b,y)^{-2}\lambda^{2}\beta(b,y)^{2}\lambda^{-1} = \lambda .\end{aligned}$$

To show (e), suppose $\tau_1 = \tau_{a,\lambda}$ and $\tau_2 = \tau_{b,\mu}$. If their lines are orthogonal, then a direct computation will show $\tau_1\tau_2 = \tau_2\tau_1$. Conversely, suppose $\tau_1\tau_2\tau_1^{-1} = \tau_2$. Thus

$$\tau_{\tau_1 b,\mu} = \tau_{b,\mu} .$$

By (d), $b = \pm\tau_1(b)$. Then $b = \pm b \pm \lambda\beta(a,b)a$. Thus, either $\lambda\beta(a,b)a = 0$ and thus $\beta(a,b) = 0$ or $2b = -\lambda\beta(a,b)a$ and likewise $\beta(a,b) = 0$ since b is a scalar multiple of a.

To show (f), suppose $\tau = \tau_{a,\lambda}$ and $\sigma a = \pm a$. Then a direct computation will give $\sigma\tau(x) = \tau\sigma(x)$ for all x in V. Conversely, if $\sigma\tau = \tau\sigma$ then $\sigma\tau\sigma^{-1} = \tau$. But $\sigma\tau\sigma^{-1} = \tau_{\sigma a,\lambda}$. By (d), $\sigma a = \pm a$.

Statement (g) is straightforward.

To show (h), suppose τ_1 and τ_2 have the same line, say $L = Ra$. By (f)

$$C(\tau_1) = \{\sigma \text{ in } Sp(V) \mid \sigma a = \pm a\} = C(\tau_2) .$$

Conversely, suppose $\tau_1 = \tau_{a,\lambda}$, $\tau_2 = \tau_{b,\mu}$ and $C(\tau_1) = C(\tau_2)$. Suppose the lines Ra and Rb are distinct. Then the hyperplanes $(Ra)^{\perp}$ and $(Rb)^{\perp}$ are distinct. Thus, there is an x in V with $\beta(a,x) = 0$ and $\beta(b,x) \neq 0$. Observe x may be chosen to be unimodular. Select a unit α in R and consider $\tau = \tau_{x,\alpha}$. Since $\tau_{x,\alpha}(a) = a + \alpha\beta(x,a)x = a$

$\tau_{x,\alpha}$ is in $C(\tau_1)$. However, we claim $\tau_{x,\alpha}(b) \neq \pm b$. Clearly, $\tau_{x,\alpha}(b) = b + \alpha\beta(x,b)x$. If $b = b + \alpha\beta(x,b)x$ then $\beta(x,b) = 0$ contradicting our choice of x . Similarly, if $-b = b + \alpha\beta(x,b)x$ then $\beta(x,b) = 0$. Hence $\tau_{x,\alpha}$ is not in $C(\tau_2)$ — again a contradiction. Hence $Ra = Rb$ and τ_1 and τ_2 have the same lines.

(C) THE CONGRUENCE SUBGROUPS OF Sp(V)

Let A be an ideal of the local ring R . In this section we define congruence subgroups of $Sp(V)$ of level A analogous to (II.A) and (III.D). In the next section we show these congruence subgroups are precisely the normal subgroups of $Sp(V)$.

The ring morphism $\Pi_A : R \to R/A$ induces a surjective R-morphism $\Pi_A : V \to V/AV$ of the symplectic spaces and a surjective group morphism $\lambda_A : Sp(V) \to Sp(V/AV)$ of their symplectic groups, e.g., see (III.21).

The _general congruence subgroup of level_ A , denoted $GSp(V,A)$, is the group

$$\lambda_A^{-1} \text{ (Center of } Sp(V/AV)\text{) ,}$$

i.e.,

$$\{\sigma \text{ in } Sp(V) \mid \lambda_A\sigma = \pm I\}$$

when $0 \neq A \neq R$. For the extreme cases, set

$$GSp(V,R) = Sp(V) \text{ and } GSp(V,0) = \{\pm I\} .$$

If $0 \neq A \neq R$ then the <u>special congruence subgroup of level</u> A , denoted $SSp(V,A)$, is the kernel of λ_A , i.e.,

$$\{\sigma \text{ in } Sp(V) \mid \lambda_A\sigma = I\} .$$

For the extreme cases, set

$$SSp(V,R) = Sp(V) \text{ and } SSp(V,0) = \{I\} .$$

As in (II), if α is in R then the <u>order</u> of α , denoted $O(\alpha)$, is the ideal generated by α .

If x is in V , the <u>order</u> of x , denoted $O(x)$, is the smallest ideal A of R satisfying

$$\Pi_A x = 0 .$$

If σ is in $Sp(V)$, the <u>order</u> of σ , denoted $O(\sigma)$, is the smallest ideal A of R satisfying

$$\lambda_A\sigma \text{ is in the Center of } Sp(V/AV) .$$

Equivalently, $\lambda_A\sigma$ is in $GSp(V,A)$.

Finally, if G is any subgroup of $Sp(V)$, then the <u>order</u> of G, denoted $O(G)$, is the smallest ideal A of R satisfying

$$G \subset GSp(V,A) .$$

For additional detail refer to (II.A).

<u>(IV.6)</u> <u>LEMMA</u>.

(a) If $A \neq R$, then

$$|GSp(V,A)/SSp(V,A)| = 2 .$$

(b) If $\tau = \tau_{a,\lambda}$ is a symplectic transvection, then $O(\tau) = O(\lambda)$.

Let A be an ideal of R and let

$T(A)$ be the group generated by symplectic transvections τ of order contained in A.

Observe, if $O(\tau) \subset A$ then $\lambda_A \tau = I$. Thus $T(A) \leq SSp(V,A)$. We show shortly that $T(A) = SSp(V,A)$.

<u>(IV.7)</u> <u>THEOREM</u>. Let $E = \{x \text{ in } V \mid x \text{ is unimodular}\}$ be the set of unimodular vectors in V. Then $T(A)$ acts on E and the orbits of E under $T(A)$ are precisely the congruence classes of E modulo A.

<u>Proof</u>. This follows from the proof of (III.23) since by (IV.2) the

isometries appearing in (III.23) may be written as a product of transvections with order contained in A.

Let T denote the family of hyperbolic planes in V. The group $T(A)$ acts on T as follows: If $H = Re \oplus Rf$ is in T and σ is in $T(A)$ then $\sigma H = \bar{H}$ where $\bar{H} = R\sigma e \oplus R\sigma f$. It is clear that $\Pi_A H = \Pi_A \bar{H}$. We show the converse of this statement.

<u>(IV.8)</u> <u>THEOREM</u>. Let the group $T(A)$ act on the family T of hyperbolic planes of V. Two hyperbolic planes H and $\bar{H}$ are in the same $T(A)$-orbit if and only if $\Pi_A H_1 = \Pi_A H_2$.

<u>Proof</u>. As above, the proof is essentially the same as (III.24).

<u>(IV.9)</u> <u>THEOREM</u>. (Characterization of $SSp(V,A)$) The group $T(A)$ generated by symplectic transvections of order contained in A is precisely $SSp(V,A)$.

<u>Proof</u>. As above, the proof is essentially the same as (III.25). However, we sketch the proof. Clearly $T(A) \subset SSp(V,A)$. Conversely, suppose σ is in $SSp(V,A)$. We show by induction on the number of hyperbolic planes of V that σ is a product of symplectic transvections of order contained in A.

Let $V = H \perp W$ where $H = Ru \oplus Rv$ is a hyperbolic plane. Then σH

is a hyperbolic plane and $\Pi_A H = \Pi_A \sigma H$. By (IV.8) there is a τ in $T(A)$ with $\tau\sigma H = H$. Then $\tau\sigma : H^{\perp} \to H^{\perp}$ ($H^{\perp} = W$). By induction there is a product δ of symplectic transvections of order contained in A in $Sp(W)$ with $\tau\sigma|_W = \delta$. Extend each transvection in δ to V by letting it be the identity on H . Then $\sigma = \tau^{-1}\delta$ is in $T(A)$.

(D) THE NORMAL SUBGROUPS OF Sp(V)

In this section we determine the normal subgroups of $Sp(V)$. In addition to the assumption that 2 is a unit in R , we assume R/m is not the field of 3 elements. The latter assumption is due to the induction technique of the proof which begins with the normal subgroups of $Sp(H)$, H a hyperbolic plane. It is shown that $Sp(H) = SL(H) = SL(V)$ ($\dim(V) = 2$). The normal subgroups of $SL(V)$ for $\dim(V) = 2$ are determined in the Appendix to this chapter. Probably the assumption on R/m could be omitted if we began by assuming $\dim(V) \geq 6$ and using the techniques of (II.D). Thus, one has either the conditions $\dim(V) \geq 2$, 2 a unit and R/m is not the field of 3 elements or $\dim(V) \geq 6$ and 2 is a unit.

If R is a field, it is well-known ([1] or [14]) that $Sp(V)/\text{Center}(Sp(V))$ is a simple group. However, if R is local one would suspect from the results in (II.D) that the normal subgroups are trapped between the congruence subgroups $GSp(V,A)$ and $SSp(V,A)$ for ideals A of R .

However, $Sp(V)$ is "smaller" than $GL(V)$ and (IV.6) gives

$$GSp(V,A)/SSp(V,A) \simeq \{I,-I\} .$$

Obviously, $Sp(V)$ is not rich in subgroups between $SSp(V,A)$ and $GSp(V,A)$. We show that the non-trivial normal subgroups of $Sp(V)$ are precisely the congruence subgroups $SSp(V,A)$ and $GSp(V,A)$ for ideals $A \neq R$ in R. The technique is that of Klingenberg [91].

(IV.10) LEMMA. Let H be a hyperbolic plane. Then

(a) $SC(H,A) = SSp(H,A)$ for an ideal A of R.

(b) In particular, $SL(H) = Sp(H)$.

(Recall, for an ideal A of R, $SC(H,A)$ was the subgroup in (II) generated by all linear transvections of order contained in A on the linear space H.)

Proof. The lemma may be verified by direct computation. Suppose (employing matrix characterization)

$$\Sigma = \begin{bmatrix} a & b \\ c & d \end{bmatrix} \quad \text{and} \quad \Gamma = \begin{bmatrix} 0 & 1 \\ -1 & 0 \end{bmatrix} .$$

If Σ is in $SSp(V,A)$, $\dim(V) = 2$, then $\Sigma\Gamma\Sigma^t = \Gamma$ and $\lambda_A\Sigma = I$. Thus, $\det(\Sigma) = 1$ and Σ is in $SC(H,A)$. Conversely, if $\det(\Sigma) = 1$ and $\lambda_A(\Sigma) = I$ then $\Sigma\Gamma\Sigma^t = \Gamma$ and Σ is in $SSp(H,A)$.

Thus, the study of $SSp(H,A)$ is no more than the theory of $SC(H,A)$. The Appendix to this chapter examines $SC(H,A)$ and $GC(H,A)$ and the normal subgroups of GL(H) for a plane H . The necessary hypothesis is that 2 is a unit and R/m is not a field of 3 elements. Under this hypothesis, if G is an SL(H)-normal subgroup of GL(H) with order the ideal A then

$$SC(H,A) \leq G \leq GC(H,A) .$$

o o

HYPOTHESIS (for remainder of chapter)

(a) 2 is a unit of R .

(b) The residue field R/m of R is not the field of 3 elements.

o o

(IV.11) THEOREM. Let $V = H \perp W$ where H is a hyperbolic plane. Let σ be in Sp(V) . Suppose

$$\sigma|_W = \text{identity} .$$

Then the normal subgroup G in Sp(V) generated by σ satisfies

$$SSp(V,O(\sigma)) \leq G \leq GSp(V,O(\sigma)) .$$

Proof. By (IV.10) and the remarks on normal subgroups of SL(H) , we have

$$SSp(H,O(\sigma|_H)) = SC(H,O(\sigma|_H))$$

contained in G. Thus G will contain all symplectic transvections whose lines lie in H and orders are contained in $O(\sigma|_H) = O(\sigma)$. But by (IV.3), $Sp(V)$ is transitive on lines and by (IV.5)

$$\beta\tau_{a,\lambda}\beta^{-1} = \tau_{\beta a,\lambda}$$

for β in $Sp(V)$ and $\tau_{a,\lambda}$ a symplectic transvection. Thus, G must contain all transvections of order $\subset O(\sigma)$.

(IV.12) COROLLARY. Let τ be a symplectic transvection. Then the normal subgroup generated by τ is $SSp(V,O(\tau))$.

The next step is to select an arbitrary ρ in $Sp(V)$ and by suitable conjugation carry it into a σ whose action is trivial except on a hyperbolic plane.

(IV.13) THEOREM. Let ρ be in $Sp(V)$. The normal subgroup G generated by ρ satisfies

$$SSp(V,O(\rho)) \leq G \leq GSp(V,O(\rho)).$$

Proof. The method is to create from ρ by conjugation an isometry whose action is restricted to a hyperbolic plane.

Let $V = H \perp W$ where $H = Ru \oplus Rv$ is a hyperbolic plane. The first basic technique is to consider commutators $\rho\tau\rho^{-1}\tau^{-1}$ where τ is a symplectic transvection rather than simply an isometry. Observe

(a) $\rho\tau\rho^{-1}\tau^{-1} = \rho(\tau\rho^{-1}\tau^{-1})$ is in G ;

(b) $\rho\tau\rho^{-1}\tau^{-1} = (\rho\tau\rho^{-1})\tau^{-1}$

$= \tau_{\rho a,\lambda}\tau_{a,-\lambda}$

if $\tau = \tau_{a,\lambda}$ by (IV.5).

This translates the problem to a study of ρa and here, the form of ρa may be suitably altered by choices of elements from $Sp(V)$. This is the second basic technique.

Let $\tau = \tau_{u,1}$. Let $A = O(\rho)$. Then

$$\bar{\rho} = \rho\tau_{u,1}\rho^{-1}\tau_{u,1}^{-1}$$
$$= \tau_{\rho u,1}\tau_{u,-1}$$

is in G and $O(\bar{\rho}) = O(\rho)$. Suppose $\rho u = \alpha u + \delta v + y$ where α,δ are in R and y is in W . Since $\lambda_A\rho = I$, $\alpha \equiv 1 \bmod A$ (and is a unit) and $O(y) \subset A$. Let σ be the <u>unitary</u> transvection

$$\sigma_{v,\alpha^{-1}y} .$$

Note

(1) $O(\sigma) = O(\alpha^{-1}y) = O(y) \subset A$.

(2) By (IV.2), σ is a product of symplectic transvections of orders contained in A .

$$(3)\quad \sigma(\rho u) = \sigma_{v,\alpha^{-1}y}(\alpha u + \delta v + y)$$
$$= \alpha u + \delta v$$

and $\sigma u = u$.

Then

$$\hat{\rho} = \sigma\tau_{\rho u,1}\sigma^{-1}\sigma\tau_{u,1}\sigma^{-1}$$
$$= \tau_{\sigma\rho u,1}\tau_{\sigma u,1}$$
$$= \tau_{\alpha u+\delta v,1}\tau_{u,1}$$

is in G , $O(\rho) = O(\hat{\rho})$ and

$$\hat{\rho}|_W = \hat{\rho}|_{H^\perp} = \text{identity} .$$

By (IV.11)

$$SSp(V,O(\rho)) \leq G .$$

Clearly $G \leq GSp(V,O(\rho))$.

We remark that the induction technique of the above proofs is based on the ideas of Klingenberg [91]. However, the use of "unitary" transvections is new and greatly simplifies Klingenberg's arguments.

(IV.14) THEOREM. (Normal Subgroup Structure of $Sp(V)$) The only proper normal subgroups of the symplectic group $Sp(V)$ are the congruence subgroups $GSp(V,A)$ and $SSp(V,A)$ for A a proper ideal of R . (We assume 2 is a unit and R/m is not the field of 3 elements.)

Proof. Let G be a normal subgroup of order $O(G) = A$. Then

$$G \leq GSp(V,A) .$$

By the previous theorem

$$SSp(V,O(\sigma)) \leq G$$

for all σ in G. But $O(G) = A$ is generated by the ideals $O(\sigma)$ for σ in G. Hence $SSp(V,A) \leq G$. Since $GSp(V,A)/SSp(V,A) \simeq \{I,-I\}$ for $A \neq R$ the theorem follows.

(IV.15) *THEOREM*. (On Commutator Subgroups)

(a) $SSp(V,A) = [Sp(V),GSp(V,A)]$

$= [Sp(V),SSp(V,A)]$.

(b) In particular,

$$Sp(V) = [Sp(V),Sp(V)] .$$

Proof. Observe the mixed commutator subgroups belong to $SSp(V,A)$. Conversely, we need to show $O([Sp(V),SSp(V,A)]) = A$ and here it is easy to see that for a given symplectic transvection τ there always is a σ in $Sp(V)$ with $O(\tau) = O(\tau^{-1}\sigma\tau\sigma^{-1})$.

(E) THE AUTOMORPHISMS OF $Sp(V)$: INTRODUCTION

In this section we assume:

(a) V is a symplectic space over a local ring R with <u>dimension</u> ≥ 6 .

(b) The residue class field R/m of R is not the field of 3 elements and 2 is a unit of R .

Let $\Lambda : Sp(V) \to Sp(V)$ be a group automorphism. The approach of this section is to examine the action of Λ on families of unimodular transvections having the same line. We show Λ carries these transvections to a family of transvections also possessing a common line. These lines are used to determine a projectivity on the projective space of V . We invoke the Fundamental Theorem of Projective Geometry (thus the projectivity is induced by a semi-linear isomorphism) and construct explicitly the automorphism Λ .

A subgroup H of a group G is <u>characteristic</u> in G if

$$\Lambda(H) \subset H$$

for all group automorphisms $\Lambda : G \to G$ of G .

The next rather simple result is crucial to our approach to the theory.

<u>(IV.16)</u> <u>THEOREM</u>. (On Characteristic Subgroups) Recall m denotes the maximal ideal of R. The subgroups $GSp(V,m)$ and $SSp(V,m)$ are characteristic subgroups of $Sp(V)$.

<u>Proof</u>. The only proper normal subgroups of $Sp(V)$ by (IV.14) are the subgroups $GSp(V,A)$ and $SSp(V,A)$ for proper ideals A of R. Hence $GSp(V,m)$ is the unique maximal normal subgroup of $Sp(V)$. Since a group automorphism will carry normal subgroups into normal subgroups, $GSp(V,m)$ will be carried into $GSp(V,m)$. By (IV.15)

$$SSp(V,m) = [Sp(V), GSp(V,m)]$$

and, consequently, $SSp(V,m)$ will be carried into itself under automorphisms of $Sp(V)$.

We use (IV.16) in the following manner: Since $SSp(V,m)$ is characteristic and

$$SSp(V,m) = \mathrm{Ker}(\lambda_m : Sp(V) \to Sp(V/mV)) ,$$

then an automorphism $\Lambda : Sp(V) \to Sp(V)$ induces naturally an automorphism

$$\Lambda_m : Sp(V/mV) \to Sp(V/mV)$$

satisfying the commutative diagram

$$(*) \qquad \begin{array}{ccc} Sp(V) & \xrightarrow{\Lambda} & Sp(V) \\ \lambda_m \downarrow & & \downarrow \lambda_m \\ Sp(V/mV) & \xrightarrow{\Lambda_m} & Sp(V/mV) \end{array}$$

where $\Lambda_m(\lambda_m \sigma) = \lambda_m(\Lambda\sigma)$.

Hence, what we prove for Λ will apply naturally to Λ_m . It will then be useful to go through the lower portion of (*) to show Λ induces a projectivity on $P(V)$.

An element σ in $Sp(V)$ is an <u>involution</u> if $\sigma^2 = I$ (identity) . With each involution σ of $Sp(V)$ (see (II.E)) there are associated unique subspaces

$$P(\sigma) = \{x \text{ in } V \mid \sigma x = x\} \qquad (\underline{\text{positive}} \text{ space of } \sigma)$$
$$N(\sigma) = \{x \text{ in } V \mid \sigma x = -x\} \qquad (\underline{\text{negative}} \text{ space of } \sigma)$$

such that V splits as the orthogonal sum

$$V = P(\sigma) \perp N(\sigma) .$$

The <u>index</u> of σ , denoted $\operatorname{ind}(\sigma)$, is defined by

$$\operatorname{ind}(\sigma) = \min\{\dim(P(\sigma)), \dim(N(\sigma))\} .$$

Observe, $\operatorname{ind}(\sigma) = \operatorname{ind}(\lambda_m\sigma)$ under $\lambda_m : Sp(V) \to Sp(V/mV)$. Further, the results of (II.E) apply since σ is in $Sp(V) \subset GL(V)$. Utilizing the techniques of (II.E) it is easy to show the next theorem.

(IV.17) THEOREM. Let

$$I_{(t)} = \{\sigma \text{ in } Sp(V) \mid \sigma \text{ is an involution of index } t\} .$$

If $\Lambda : Sp(V) \to Sp(V)$ is a group automorphism, then

$$\Lambda I_{(t)} = I_{(t)} .$$

Suppose an involution σ in $Sp(V)$ determines a splitting $\sigma : P \perp N$ of V . If β is in $Sp(V)$ then $\beta\sigma\beta^{-1}$ is an involution with spaces βP and βN . Further, $\beta\sigma = \sigma\beta$ if and only if $\beta N \subseteq N$ and $\beta P \subseteq P$. Induction on n will prove the following theorem.

(IV.18) THEOREM. A collection X of involutions in $Sp(V)$ is pairwise commuting if and only if V splits into hyperbolic planes

$$V = H_1 \perp \cdots \perp H_m$$

such that each involution σ in X has the form

$$\sigma|_{H_i} = \pm \text{ (identity on } H_i \text{)}$$

for $1 \leq i \leq m$.

(F) PRESERVATION OF TRANSVECTIONS

Let $\tau = \tau_{a,\lambda}$ be a symplectic transvection. Recall the order of τ, $O(\tau) = R\lambda$, and τ is *unimodular* if $O(\tau) = R$, i.e., λ is a unit.

The purpose of this section is to provide a proof of the next result. We continue the assumptions of (E).

(IV.19) THEOREM. Let $\Lambda : Sp(V) \to Sp(V)$ be a group automorphism. If τ is a unimodular transvection, then $\Lambda\tau$ is a unimodular transvection.

Proof. Let $\tau = \tau_{a,\lambda}$ be a unimodular transvection. Since the vector a is unimodular, there is a hyperbolic basis V, say

$$\{a = x_1, y_1, \ldots, x_m, y_m\},$$

with

$$V = P_1 \perp \cdots \perp P_m$$

where $P_i = Rx_i \oplus Ry_i$ for $1 \le i \le m$.

Thus the line $L = Ra$ of τ is in P_1. Let $T = \Lambda\tau$.

Choose σ to be an involution of index 2 whose plane contains L, say

$$\sigma = -i_{P_1} \perp i_{P_2} \perp \cdots \perp i_{P_r}.$$

Let $\sum = \Lambda\sigma$. By (IV.17) $\sum$ is an involution of index 2 and thus $\sum$ determines an orthogonal splitting of V

$$\sum : V = P \perp N$$

with, say, $\dim(P) = 2$. Further τ and σ permute and hence $\sum T = T\sum$. This implies $TP \subseteq P$ and $TN \subseteq N$.

The aim is to show T is a unimodular transvection. We begin by showing T^2 is the identity on N . This will involve several steps.

Select an involution $\bar{\sum}$ ($\neq \pm \sum$) (since $n \geq 6$) of index 2 which permutes with $\sum$. Then, similarly, $\bar{\sum}$ determines a splitting

$$\bar{\sum} : V = \bar{P} \perp \bar{N}$$

with plane $\bar{P}$ (we are employing P and $\bar{P}$ to denote planes rather than the "positive space" of the last section). We claim

$$T\bar{P} = \bar{P} .$$

To show this, note Λ^{-1} is an automorphism of $Sp(V)$; thus $\Lambda^{-1}\bar{\sum}$ is an involution of index 2 , $\Lambda^{-1}\bar{\sum} \neq \pm\sigma$, and $(\Lambda^{-1}\bar{\sum})\sigma = \sigma(\Lambda^{-1}\bar{\sum})$. By (IV.18) there is a splitting of V into hyperbolic planes,

$$V = Q_1 \perp \cdots \perp Q_m ,$$

where $\sigma|_{Q_j} = \pm i_{Q_j}$ and $(\Lambda^{-1}\bar{\sum})|_{Q_j} = \pm i_{Q_j}$ for $1 \leq j \leq m$. Without loss we may assume the plane of σ is $Q_1 = P_1$ and the plane of $\Lambda^{-1}\bar{\Sigma}$ is Q_2, i.e., their planes are orthogonal. Then

$$\tau(\Lambda^{-1}\bar{\Sigma}) = (\Lambda^{-1}\bar{\Sigma})\tau$$

and hence

$$T\bar{\Sigma} = \bar{\Sigma}T .$$

Thus $T\bar{P} \subseteq \bar{P}$ and $T\bar{N} \subseteq \bar{N}$. Since T is an isometry $T\bar{P} = \bar{P}$ and $T\bar{N} = \bar{N}$.

We next show $T = \pm i_{\bar{P}}$.

Suppose $\bar{P}$ is spanned by the hyperbolic pair $\{v,w\}$. Recall $\bar{P} \subseteq N$. Consider the line $\bar{L} = Rv$ (similarly $\bar{L} = Rw$). Since $n \geq 6$, there is an element u in the hyperbolic basis for V with u in $(P \perp \bar{P})^{\perp}$. Let $J = R(u + v)$. Observe

(a) $J \subseteq N$,

(b) $J \perp \bar{L}$,

(c) $J \not\subseteq \bar{P}$,

(d) $J \not\perp \bar{P}$.

Let ϕ be a unimodular transvection in $Sp(V)$ having line J. Set $\Sigma^* = \phi\Sigma\phi^{-1}$. Then Σ^* is an involution of index 2, $\Sigma^* \neq \pm\Sigma$, and the plane of Σ^* is $P^* = \phi P$. As in the above paragraph, we conclude $TP^* = P^*$. Thus $T(\bar{P} \cap P^*) = \bar{P} \cap P^*$ and consequently $T\bar{L} = \bar{L}$. That is, $T(Rv) = Rv$ (similarly $T(Rw) = Rw$). Since T is R-linear there exist r_v and r_w in R with

$$T(x) = r_v x \ , \quad T(y) = r_w y$$

for all x in Rv and y in Rw . Since $\{v + w, w\}$ is also a hyperbolic pair spanning $\bar{P}$, the above argument implies there is an r_{v+w} in R with

$$T(z) = r_{v+w} z$$

for all z in $R(v + w)$. Then

$$\begin{aligned} r_{v+w}(v + w) &= T(v + w) \\ &= Tv + Tw \\ &= r_v v + r_w w \end{aligned}$$

and thus

$$r_v = r_w = r_{v+w} = r \ .$$

Then $1 = \beta(v,w) = \beta(Tv,Tw) = \beta(rv,rw) = r^2$. Hence $r = \pm 1$ and

$$T|_{\bar{P}} = \pm i_{\bar{P}} \ .$$

Let $\sigma_j = (i_{P_1}) \perp \cdots \perp (i_{P_{j-1}}) \perp (-i_{P_j}) \perp (i_{P_{j+1}}) \perp \cdots \perp (i_{P_m})$ for $1 \le j \le m$ and let $S = \{\sigma_1, \sigma_2, \ldots, \sigma_m\}$. Let $\bar{P}_1, \ldots, \bar{P}_m$ be the planes of $\Lambda\sigma_1, \ldots, \Lambda\sigma_m$, respectively. Applying the above argument with $\bar{\Sigma}$ being each of $\Lambda\sigma_i$, $i = 2, \ldots, m$, conclude

$$T|_{\bar{P}_j} = \pm i_{\bar{P}_j} , \quad 2 \le j \le m .$$

Hence T^2 acts as the identity on

$$N = \bar{P}_2 \perp \cdots \perp \bar{P}_m .$$

The plane P_1 has a hyperbolic basis $\{x_1, y_1\}$ and the plane P_2 has hyperbolic basis $\{x_2, y_2\}$. Let $\bar{J}$ be the line $\bar{J} = R(x_1 + x_2)$. Then

(a) $\bar{J} \perp L$ since $L = Rx_1$,

(b) $\bar{J} \not\subseteq P_1$,

(c) $\bar{J} \not\perp P_1$.

Let δ be a unimodular transvection with line $\bar{J}$. Let $\sigma_0 = \delta\sigma\delta^{-1}$. Then (a) σ_0 is an involution of index 2 ,

(b) the plane of σ_0 is $P_0 = \delta P$,

(c) $L \subseteq P_0$,

(d) $\sigma_0 \neq \pm\sigma$.

Thus σ and σ_0 are distinct involutions of index 2 with proper plane containing L . We have $\sum = \Lambda\sigma$. Let $\sum_0 = \Lambda\sigma_0$. Then $\sum$ and $\sum_0$ determine splittings of V

$$\textstyle\sum : V = P \perp N$$

and

$$\textstyle\sum_0 : V = P_0 \perp N_0 .$$

Since 2 is a unit of R , set $\bar{T} = \Lambda\tau_{a,\lambda/2}$. Thus, by the above

$$
\begin{aligned}
(\bar{T})^2 &= \Lambda\tau_{a,\lambda/2}\Lambda\tau_{a,\lambda/2} \\
&= \Lambda\tau_{a,\lambda/2+\lambda/2} \\
&= \Lambda\tau_{a,\lambda}
\end{aligned}
$$

acts as the identity on N and N_0 .

Let $\Pi_m : N \to \Pi_m N$ and $\Pi_m : N_0 \to \Pi_m N_0$ under $\Pi_m : V \to V/mV$. Apply

$$\lambda_m : Sp(V) \to Sp(V/mV)$$

to σ , σ_0 and δ_0 and repeat the above arguments. One concludes that the k-spaces $\Pi_m N$ and $\Pi_m N_0$ of dimension $n - 2$ are distinct. Since $\Pi_m N \neq \Pi_m N_0$, there is an $\bar{x} \neq \bar{0}$ in $\Pi_m N_0 - \Pi_m N$. Thus there is a unimodular x in $N_0 - N$. Let $\{\bar{b}_1,\ldots,\bar{b}_{n-2}\}$ be a k-basis for $\Pi_m \bar{N}$ obtained from an R-basis $\{b_1,\ldots,b_{n-2}\}$ of N . Then for some $\bar{y}$ in V/mV , $\{\bar{b}_1,\ldots,\bar{b}_{n-2},\bar{x},\bar{y}\}$ is a basis for V/mV . Let $\Pi_m y = \bar{y}$. Then $\{b_1,\ldots,b_{n-2},x,y\}$ is an R-basis for V . But $N = Rb_1 \oplus \cdots \oplus Rb_{n-2}$ and Rx is in N_0 . Since $\Lambda\tau_{a,\lambda}$ fixes N and N_0 , $\Lambda\tau_{a,\lambda}$ is the identity on the hyperplane

$$H = N \oplus Rx \ .$$

That is, $\Lambda\tau_{a,\lambda}$ is a transvection. Apply diagram (*) following (IV.16) to Λ and $\tau_{a,\lambda}$ and conclude $\bar{\Lambda}(\lambda_m\tau_{a,\lambda})$ is non-trivial. Hence $\Lambda\tau_{a,\lambda}$ is unimodular.

(IV.20) COROLLARY. Let Λ be an automorphism of Sp(V) and τ a symplectic transvection. Then $\Lambda\tau$ is a symplectic transvection.

Proof. Let $\tau = \tau_{a,\lambda}$ be a symplectic transvection. If λ is a unit the result follows from (IV.19). Suppose λ is not a unit. Then

$$\tau_{a,\lambda} = \tau_{a,1}\tau_{a,\lambda-1}$$

where 1 and $\lambda - 1$ are units. Hence $\tau_{a,1}$ and $\tau_{a,\lambda-1}$ are unimodular and by (IV.19), $T_1 = \Lambda\tau_{a,1}$ and $T_2 = \Lambda\tau_{a,\lambda-1}$ are unimodular symplectic transvections. For σ in Sp(V), let $C(\sigma) = \{\beta$ in Sp(V) $\mid \sigma\beta = \beta\sigma\}$ be the centralizer of σ. By (IV.5)(h) $C(\tau_{a,1}) = C(\tau_{a,\lambda-1})$. Then $C(T_1) = C(T_2)$. Since T_1 and T_2 are unimodular transvections, by (IV.5)(h), they have the same line. Hence $\Lambda\tau = T_1T_2$ is a transvection.

(G) THE AUTOMORPHISMS OF Sp(V)

We will continue the hypothesis of Section (E).

Suppose $g : V \to V$ is a μ-semi-linear isomorphism of the R-space V onto itself. We say g preserves orthogonality if for all unimodular x and y, $\beta(x,y) = 0$ implies $\beta(gx,gy) = 0$.

(IV.21) PROPOSITION. The semi-linear isomorphism $g : V \to V$ preserves orthogonality if and only if there is a unit α in R with

$$\beta(gx,gy) = \alpha(\beta(x,y))^{\mu}$$

for all x and y in V where $\mu : R \to R$ is the ring automorphism associated with g.

Proof. Suppose g preserves orthogonality. Let $B = \{e_1,f_1,e_2,f_2,\ldots, e_m,f_m\}$ be a hyperbolic basis for V. For $1 \le i \le m$ set $\alpha_i = \beta(ge_i,gf_i)$. Then

$$\beta(e_i + e_j,f_i - f_j) = 0$$

for $1 \le j \le m$ implies

$$\beta(ge_i + ge_j,gf_i - gf_j) = 0$$

and consequently $\alpha_i = \alpha_j$. Hence, if $\alpha = \alpha_i$ then

$$\beta(ge_i,gf_j) = \alpha\delta_{ij}$$

and consequently

$$\beta(gx_i,gx_j) = \alpha\beta(x_i,x_j)$$

for all x_i and x_j from B. Then by linearity

$$\beta(gx,gy) = \alpha(\beta(x,y))^{\mu}$$

for all x and y in V (note for δ in R, $\delta^{\mu} = \mu(\delta)$). It remains to show α is a unit. Since (V,β) is an inner product space, the map

$$\beta^d : V \to V^*$$

by

$$\beta^d(x) = \beta(x, \ )$$

is an R-isomorphism; since g is a semi-linear isomorphism, ge_1 is unimodular. Hence $\beta^d(ge_1) = \beta(ge_1, \ ) : V \to R$ is surjective, i.e., there is a gx with

$$\beta(ge_1, gx) = 1$$

(note each y in V has the form gx for some x in V). That is,

$$\alpha(\beta(e_1, x))^\mu = 1$$

and α is a unit.

(IV.22) COROLLARY. Suppose a semi-linear isomorphism $g : V \to V$ preserves orthogonality. Then

(a) g^{-1} preserves orthogonality.

(b) For all x and y in V, $\beta(x,y) = 0$ if and only if $\beta(gx,gx) = 0$.

Let $g : V \to V$ be a semi-linear isomorphism of V. In (II.F) we defined a map

$$\Phi_g : GL(V) \to GL(V)$$

by

$$\Phi_g(\sigma) = g\sigma g^{-1}$$

for all σ in GL(V) . The basic properties of Φ_g were given in (II.F). We seek to describe the automorphisms of Sp(V) in terms of the Φ_g . Thus, the first question is, "For what g does

$$\Phi_g : Sp(V) \to Sp(V) \text{ ?"}$$

<u>(IV.23)</u> <u>PROPOSITION</u>. The automorphism $\Phi_g : GL(V) \to GL(V)$ induces an automorphism on Sp(V) if and only if g preserves orthogonality.

<u>Proof</u>. Suppose first that g preserves orthogonality and σ is in Sp(V) . Then

$$\begin{aligned}
\beta(\Phi_g\sigma x, \Phi_g\sigma y) &= \beta(g\sigma g^{-1}x, g\sigma g^{-1}y) \\
&= \alpha[\beta(\sigma g^{-1}x, \sigma g^{-1}y)]^{\mu} \\
&= \alpha[\beta(g^{-1}x, g^{-1}y)]^{\mu} \\
&= \beta(x,y)
\end{aligned}$$

and $\Phi_g\sigma$ is in Sp(V) . Conversely, suppose $\Phi_g\sigma$ is in Sp(V) for each σ in Sp(V) . Let x and y be unimodular orthogonal vectors in V . Let τ_1, τ_2 be symplectic transvections with lines Rx and Ry . Then by (IV.5) τ_1 and τ_2 commute. Hence the symplectic unimodular transvections $\Phi_g\tau_1 = g\tau_1 g^{-1}$ and $\Phi_g\tau_2 = g\tau_2 g^{-1}$ with lines Rgx and Rgy also commute. Again by (IV.5), $\beta(gx, gy) = 0$.

(IV.24) PROPOSITION. Let $g : V \to V$ be a semi-linear isomorphism which preserves orthogonality. If $\tau = \tau_{a,\lambda}$ is a symplectic transvection, then

$$\Phi_g(\tau) = \tau_{ga,\mu(\lambda)\alpha^{-1}}$$

where μ and α are as above, i.e., see (IV.21).

An automorphism $\Lambda : Sp(V) \to Sp(V)$ is called radial if there is a group morphism

$$\chi : Sp(V) \to \mathrm{Center}(Sp(V)) = \{\pm I\}$$

such that

$$\Lambda(\sigma) = \chi(\sigma)\sigma$$

for all σ in $Sp(V)$ (see (II.F)). A radial automorphism Λ determines a unique χ and the automorphism Λ is written in the form

$$P_\chi .$$

The next two propositions are straightforward.

(IV.25) PROPOSITION. Let χ be any morphism $\chi : Sp(V) \to \mathrm{Center}(Sp(V))$. Then χ determines a radial automorphism P_χ if and only if $\chi(\pm I) = I$.

(IV.26) PROPOSITION. (Uniqueness) Suppose P_{χ_1} , P_{χ_2} , Φ_{g_1} and Φ_{g_2} are automorphisms of $Sp(V)$ as described above. The following are equivalent:

(a) $P_{\chi_1} \circ \Phi_{g_1} = P_{\chi_2} \circ \Phi_{g_2}$.

(b) $P_{\chi_1} = P_{\chi_2}$ and $\Phi_{g_1} = \Phi_{g_2}$.

(c) $\chi_1 = \chi_2$ and $g_1 = \alpha g_2$ for α a unit in R .

We next initiate the proof that if $\Lambda : Sp(V) \to Sp(V)$ is a group automorphism, then there is a χ and g with

$$\Lambda = P_\chi \circ \Phi_g \ .$$

In contrast to (II.24), the symplectic group will possess only automorphisms of the first type. This simplifies the theory.

(IV.27) LEMMA. Let Γ be an automorphism of $Sp(V)$ such that for each symplectic transvection τ in $Sp(V)$, $\Gamma\tau$ is a transvection having the same line as τ . Then

$$\Gamma = P_\chi$$

for $\chi : Sp(V) \to \mathrm{Center}(Sp(V))$ a group morphism.

Proof. Let σ be in $Sp(V)$, L a line in V and τ a unimodular symplectic transvection with line L . Then

$$(\Gamma\sigma)(\Gamma\tau)(\Gamma\sigma)^{-1}$$

is a transvection with line $(\Gamma\sigma)L$, while $\Gamma(\sigma\tau\sigma^{-1})$ is a transvection with line σL . Hence $(\Gamma\sigma)L = \sigma L$ and thus

$$(\sigma^{-1}\Gamma\sigma)L = L .$$

Let $L = Re$. Then

$$(\sigma^{-1}\Gamma\sigma)(e) = r_e e \qquad (r_e \text{ in } R).$$

Similarly, if f is unimodular,

$$(\sigma^{-1}\Gamma\sigma)(f) = r_f f \qquad (r_f \text{ in } R).$$

If e and f are chosen R-free, then $e + f$ is unimodular. Consequently,

$$r_{e+f}(e + f) = r_e e + r_f f$$

and $r_{e+f} = r_e = r_f = r$. Thus, extending linearly,

$$(\sigma^{-1}\Gamma\sigma)x = rx$$

for x in V . Taking x and y with $\beta(x,y) = 1$ we have $1 = \beta(x,y) = \beta(\sigma^{-1}\Gamma\sigma x,\sigma^{-1}\Gamma\sigma y) = \beta(rx,ry) = r^2$ and $r = \pm 1$. Hence

$$\Gamma\sigma = \pm\sigma .$$

We begin the general case.

Let L be a line in V and

$$\Lambda : Sp(V) \to Sp(V)$$

a group automorphism.

Let $T(L) = \{\tau \mid \tau$ is a unimodular transvection with line $L\}$.

Then each element of $\Lambda(T(L))$ is a unimodular symplectic transvection. Let τ_1 and τ_2 be in $T(L)$ and $T_1 = \Lambda\tau_1$ and $T_2 = \Lambda\tau_2$. Since τ_1 and τ_2 have the same line then their centralizers in $Sp(V)$, i.e.,

$$\begin{aligned} C(\tau_1) &= \{\sigma \text{ in } Sp(V) \mid \sigma\tau_1 = \tau_1\sigma\} \\ &= \{\sigma \text{ in } Sp(V) \mid \sigma\tau_2 = \tau_2\sigma\} \\ &= C(\tau_2) \end{aligned}$$

are equal by (IV.5)(h). Thus

$$C(T_1) = C(T_2)$$

and, again by (IV.5)(h), T_1 and T_2 have the same line. Thus, to L we assoicate a line $L^\wedge$ in V when

$$\Lambda(T(L)) \subseteq T(L^\wedge) .$$

Since Λ^{-1} is a group automorphism, $\Lambda(T(L)) = T(L^\wedge)$. Let $P(V)$ denote the projective space of V , i.e., the collection of lines in V (see (I.D) or (II.F)).

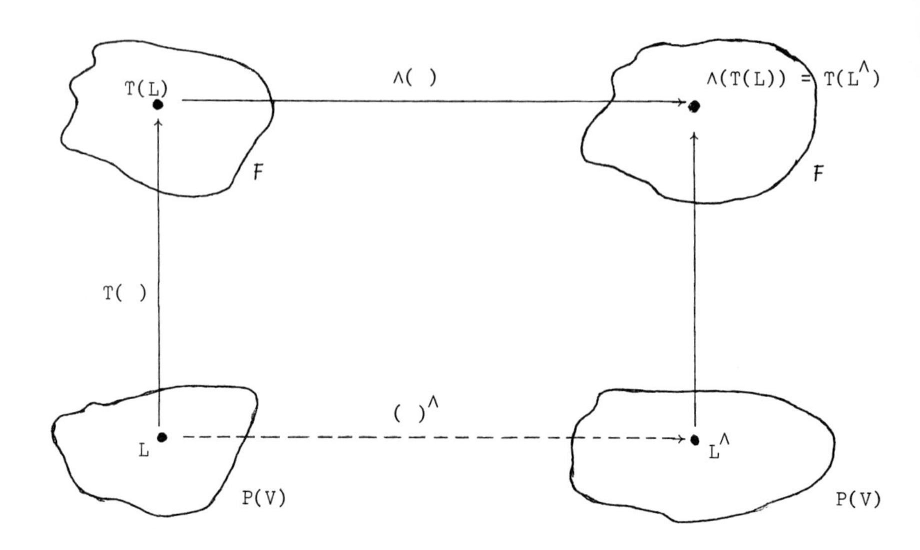

F = {T(L) | T(L) = family of all symplectic unimodular transvections with line L as L varies through P(V)}

The above association $L \to L^\Lambda$ determines a bijection between $P(V)$ and $P(V)$. Denote the map $L \to L^\Lambda$ by α. The bijection $\alpha : P(V) \to P(V)$ will carry orthogonal lines to orthogonal lines. For suppose $\beta(L_1,L_2) = 0$.[1] Select unimodular transvections τ_1 and τ_2 with lines L_1 and L_2, respectively. Since $\beta(L_1,L_2) = 0$, τ_1 and τ_2 commute. Hence $\Lambda\tau_1$ and $\Lambda\tau_2$ commute which implies $\alpha L_1 = L_1^\Lambda$ and $\alpha L_2 = L_2^\Lambda$ are orthogonal. Similarly, α^{-1} preserves orthogonality.

Thus, $\alpha : P(V) \to P(V)$ is a bijective mapping preserving orthogonal lines. We want to show α is a projectivity.

---

[1] For this notation see following paragraphs.

A difficulty encountered at this point is that there may be elements in V which are not contained in any line of V. Example (b), following (I.6), illustrates this situation.

To handle this problem the following notation is introduced.

Let W be a subset of V and Q a subset of $P(V)$ ($P(V)$ = collection of lines in V). If L and $\bar{L}$ are in $P(V)$, write $L \perp \bar{L}$ or $\beta(L,\bar{L}) = 0$ if

$$\beta(x,y) = 0 \quad \text{for all } x \text{ in } L \text{ and } y \text{ in } \bar{L} .$$

Recall, if $L \perp \bar{L}$, the line L is said to be orthogonal to the line $\bar{L}$.

Let

$$P(W) = \{L \text{ in } P(V) \mid L \subset W\} ,$$

$$W^{\perp} = \{x \text{ in } V \mid \beta(W,x) = 0\} ,$$

$$\langle W\rangle = R\text{-submodule of } V \text{ generated by } W .$$

Let

$$Q^{0} = \{L \text{ in } P(V) \mid L \perp \bar{L} \text{ for all } \bar{L} \text{ in } Q\}$$

$$\langle Q\rangle = R\text{-submodule generated by all vectors } x \text{ where } x \text{ is in a line belonging to } Q .$$

Observe

(a) $\langle P(W)\rangle \subset \langle W\rangle$

(b) $Q \subset P(\langle Q\rangle)$

(c) If W is a subspace of V then

(i) $W = \langle W \rangle = \langle P(W) \rangle$

(ii) $P(W^{\perp}) = (P(W))^0$

(iii) $\langle P(W^{\perp}) \rangle = \langle W^{\perp} \rangle = W^{\perp}$.

If W is a subspace of V, let $\alpha P(W) = \{\alpha L \mid L$ is a line of $W\}$.

Let H be any hyperplane in V. Let $L = H^{\perp}$. Then $H^{\perp} = L$ is a line and $H^{\perp\perp} = L^{\perp} = H$.

Also

$$L = (P(H))^0 .$$

If $\bar{L}$ is any line in $P(H)$, $\bar{L} \perp L$. Thus $\alpha\bar{L} \perp \alpha L$ and

$$\alpha L = [\alpha(P(H))]^0 .$$

Thus

$$(\alpha L)^0 = \alpha(P(H))$$

and

$$(\alpha L)^{\perp} = \langle \alpha(P(H)) \rangle$$

is a hyperplane. We conclude that if H is a hyperplane then $\langle \alpha P(H) \rangle$ is a hyperplane which we denote by $\bar{H}$.

and if $H^{\perp} = L$ then $\bar{H} = (\alpha L)^{\perp}$; equivalently, $P(H)^0 = L$ then $P(\bar{H})^0 = \alpha L$. Finally, $P(\bar{H}) = \alpha P(H)$.

<u>(IV.28)</u> <u>LEMMA</u>. The bijection α carries planes to planes.

<u>Proof</u>. Recall, by (IV.16), $\Lambda : Sp(V) \to Sp(V)$ induces an automorphism $\Lambda_m : Sp(V/mV) \to Sp(V/mV)$ satisfying the commutative diagram

$$(*) \qquad \begin{array}{ccc} Sp(V) & \xrightarrow{\Lambda} & Sp(V) \\ \lambda_m \downarrow & & \downarrow \lambda_m \\ Sp(V/mV) & \xrightarrow{\Lambda_m} & Sp(V/mV) \end{array} \quad .$$

In turn, the bijection $\alpha : L \to \alpha L$ induces a bijection

$$\alpha_m : P(V/mV) \to P(V/mV)$$

by

$$\alpha_m : \Pi_m L \to \Pi_m(\alpha L)$$

such that

$$(**) \qquad \begin{array}{ccc} P(V) & \xrightarrow{\alpha} & P(V) \\ \Pi_m \downarrow & & \downarrow \Pi_m \\ P(V/mV) & \xrightarrow{\alpha_m} & P(V/mV) \end{array}$$

is commutative.

Suppose $W = L_1 \oplus L_2$ is a plane and $Q = W^{\perp}$, $H_1 = L_1^{\perp}$ and $H_2 = L_2^{\perp}$. Then

$$Q = W^{\perp} = (L_1 \oplus L_2)^{\perp}$$
$$= L_1^{\perp} \cap L_2^{\perp}$$
$$= H_1 \cap H_2 \quad .$$

Further, Q is a space of dimension $n - 2$.

Also

$$\alpha P(Q) = P(W)^0 = L_1^0 \cap L_2^0$$
$$= P(H_1) \cap P(H_2) \; .$$

Since α is bijective

$$P(Q) = \alpha(P(H_1) \cap P(H_2))$$
$$= \alpha P(H_1) \cap \alpha P(H_2) \; .$$

Let $\bar{H}_1 = \langle \alpha P(H_1) \rangle$, $\bar{H}_2 = \langle \alpha P(H_2) \rangle$. Then, $\bar{H}_1$ and $\bar{H}_2$ are hyperplanes. Further, by checking (**) $\Pi_m \bar{H}_1 \neq \Pi_m \bar{H}_2$. Hence, by (I.7), $\bar{H}_1 \cap \bar{H}_2$ is an R-space of dimension $n - 2$. Each basis element b in $\bar{H}_1 \cap \bar{H}_2$ gives rise to a line Rb in both $P(\bar{H}_1) = \alpha P(H_1)$ and $P(\bar{H}_2) = \alpha P(H_2)$. Hence Rb is in $\alpha P(Q)$ and b is in $\langle \alpha P(Q) \rangle$. Hence

$$\langle \alpha P(Q) \rangle = H_1 \cap H_2$$

and $\langle \alpha P(Q) \rangle$ is a space of dimension $n - 2$. But $\langle \alpha P(Q) \rangle^{\perp} = \langle \alpha P(W) \rangle$ is thus a space of dimension 2 . By checking (**), αL_1 and αL_2 are independent and we have

$$\langle \alpha(P(W)) \rangle = \alpha L_1 \oplus \alpha L_2 \; .$$

<u>(IV.29)</u> <u>THEOREM</u>. The bijective map $\alpha : P(V) \to P(V)$ is a projectivity.

<u>Proof</u>. This follows from the above discussion.

By the Fundamental Theorem of Projective Geometry (I.13), there is a semi-linear isomorphism $g : V \to V$ satisfying

$$\alpha(Ra) = Rg(a)$$

for each line $L = Ra$ in $P(V)$. Further, since α preserves orthogonal lines, g preserves orthogonality.

Thus, for the automorphism $\Lambda : Sp(V) \to Sp(V)$ the induced projectivity

$$\alpha : L \to \alpha L = L^{\Lambda}$$

is given by the semi-linear isomorphism g.

Consider the automorphism

$$\Phi_g^{-1} \circ \Lambda$$

of $Sp(V)$. Let τ be a symplectic transvection with line L. Then $\Lambda\tau$ is a transvection with line $\bar{L} = \alpha L = g(L)$. Since $\Phi_g^{-1} \circ \Lambda = \Phi_{g^{-1}} \circ \Lambda$, $(\Phi_g^{-1} \circ \Lambda)(\tau) = \Phi_{g^{-1}}(\Lambda\tau)$ is a transvection with line $g^{-1}(\bar{L}) = g^{-1}(gL) = L$. Hence, the line of $(\Phi_g^{-1} \circ \Lambda)(\tau)$ is the same as τ. By (IV.27), $\Phi_g^{-1} \circ \Lambda = \Gamma$ where $\Gamma = P_\chi$ for a group morphism $\chi : Sp(V) \to \{\pm I\}$. Therefore, $\Lambda : \Phi_g \circ \Gamma$. Using Λ^{-1} instead of Λ we get

$$\Lambda = P_\chi \circ \Phi_g .$$

(IV.30) THEOREM. (Automorphisms of $Sp(V)$) Let V be a symplectic space over a local ring R of dimension ≥ 6 . Assume 2 is a unit in R and R/m is not the field of 3 elements. If $\Lambda : Sp(V) \to Sp(V)$ is an automorphism of $Sp(V)$, then there exist automorphisms P_χ and Φ_g with

$$\Lambda : P_\chi \circ \Phi_g .$$

(H) DISCUSSION

An examination of $Sp(V)$ when R is a field is given by Dieudonné [14], [63], [64] and Artin [1]. One should also note Siegel's famous paper in 1943 [133] on symplectic geometry. When R is local often a parody of the field techniques will carry the theory through. If R is a domain, basic results are more difficult for the subgroup theory as $GL(V)$ might indicate. A general formulation of the normal subgroups of the symplectic (and split) group is given by Bak in [34] and [35] in finite dimensions, while a discussion of the stable symplectic (and split) group $Sp(R) = \varinjlim Sp_n(R)$ and its normal subgroups occurs in Bass [8]. In the above cases, the normal subgroups are trapped between "congruence" subgroups in "stable" situations. A discussion of the symplectic group over a local ring <u>without</u> assumption that the form is non-singular was given by Riehm [122] in 1966. In Riehm's paper the local ring must be replaced by a valuation ring. In this setting there exist normal subgroups which are not congruence subgroups; however, Riehm obtains a complete description of the normal subgroups. Recently Chang [53] extended the non-singular case to semi-local domains.

One observes the symplectic transvection is a convenient generator of $Sp(V)$. Concerning the symplectic transvections — much of the study of the classical linear groups is founded on the idea that they are generated by "elementary" transformations which fixed a hyperplane elementwise. Dieudonné in [64] describes the generators of the classical linear groups over a field and discusses the minimal number of generators needed to represent a given element.

As in the previous chapters, the initial concern is the action of $Sp(V)$ and its congruence subgroups on the set of unimodular vectors of V. This is immediately useful in determining the normal subgroups; e.g., take σ in a normal subgroup G and $\tau_{a,\lambda}$ a symplectic transvection. On one hand $\sigma\tau\sigma^{-1}\tau^{-1}$ is in G while on the other hand

$$\sigma\tau_{a,\lambda}\sigma^{-1}\tau_{a,\lambda}^{-1} = \tau_{\sigma a,\lambda}\tau_{a,\lambda}^{-1} ,$$

and the element is determined by the action of σ on a . Hence, the action of Sp(V) and its congruence subgroups on unimodular elements allows commutators to be conveniently described.

Symplectic methods have proven useful in work toward a solution of Serre's Conjecture, i.e., projective modules over polynomial rings over fields are free. (See Swan [142], [17] or Bass [39] and (I.G).)

Since about 1948 the automorphisms of the symplectic group over various rings have been investigated by Hua [80], Dieudonné [63], Rickart [120], Reiner [117], Wan and Wang [151], O'Meara [110], Keenan [88] and McDonald and McQueen [103]. Reiner determined the automorphisms of Sp(V) over the ring of rational integers; Wan announced in [150] that he had determined the automorphisms of Sp(V) over a Euclidean domain; O'Meara determined the automorphisms of Sp(V) over any domain of any characteristic when $\dim(V) \geq 4$; Keenan, McQueen and McDonald examined the automorphisms over local rings. Central to these approaches were methods involving centralizers and involutions. Recently Solazzi [134] examined the automorphisms of the symplectic congruence subgroups over domains, employing the involution-free methods of O'Meara [23]. Solazzi's approach was motivated by the fact that symplectic congruence subgroups need not contain any non-trivial involutions and hence the usual techniques fail. In 1973, A. Hahn [72] gave a uniform isomorphism theory for certain subgroups of the isometry groups. Among Hahn's results is an isomorphism theory for the symplectic group (also the linear and unitary groups) over a domain independent of the characteristic of the domain. Another classical difficulty in automorphism theory is that a method of involutions does not readily apply to projective symplectic groups — here one encounters the problem of distinguishing in a group theoretic fashion between projective involutions of the "first" or "second" type. This problem was also remedied by Solazzi. We have not attempted to examine the automorphisms of the congruence subgroups in this monograph. This has been done for domains but at this writing not for local rings. However, one suspects that the automorphisms of the congruence subgroups are also of the standard form and, further, induce the automorphisms of $Sp(V)$ (similarly for the general linear and orthogonal groups).

Recently, O'Meara [113] completed a thorough examination of the symplectic group and its automorphisms over a domain. For domains, O'Meara reduced the previous lower bound on $\dim(V)$ and extended the theory to collinear transformations (as he did for $GL(V)$ in [23]).

A word about collinear transformations: A collinear transformation of a k-vector space V (k a field) onto itself is a semi-linear isomorphism. If $SGL(V)$ denotes the group of collinear transformations, i.e., the group of invertible semi-linear isomorphisms of V (see Section (I.E)), then $GL(V)$ is a normal subgroup of $SGL(V)$. If a group automorphism $\Lambda : SGL(V) \to SGL(V)$, then [23] $\Lambda = P_\chi \circ \Phi_g$ or $\Lambda = P_\chi \circ \Psi_h$ and if we move to the projective collinear transformations the P_χ vanishes — hence the automorphisms are essentially inner and automorphisms of GL(V) are restrictions of certain automorphisms of SGL(V) . It can be argued that the automorphism theory more appropriately should be formulated for SGL(V) and its corresponding projective group. Since this monograph was thought of as an introduction we hesitated to develop the more general, though probably correct, context.

Further, concerning automorphisms, most approaches are set in finite dimensional spaces. However, usually only "codimension" or "dimension of residue spaces" of the elements of GL(V) , Sp(V) , etc. are needed. Thus, most of the above treatments carry over to infinite dimensions with modest changes.

As occurred for GL(V) , the general theory for Sp(V) for arbitrary rings has moved in two directions. To examine internal and normal subgroup theory for Sp(V) generally the dim(V) has been pushed up, increasing freedom of movement of vectors, lines, planes, etc. and surpassing the dimension of the maximal spectrum until a standard "stable" theory is achieved. To examine Aut(Sp(V)) (V is an R-space of fixed dimension), the ring R is generally embedded as a subring in a ring S having all projective modules free, then involutions, etc., split the extended S-space $S \otimes_R V$ and have associated free summands.

As noted already, if dim(V) = 2 then Sp(V) = SL(V) . We remarked on GL(V) and SL(V) when dim(V) = 2 in (II).

APPENDIX: NORMAL SUBGROUPS OF GL(V) : dim(V) = 2

In this chapter the normal subgroups of Sp(V) were determined by an induction technique beginning with dim(V) = 2 . If dim(V) = 2 then (IV.10) shows Sp(V) is precisely SL(V) . Hence, in this section we examine SL(V) and GL(V) when dim(V) = 2 . Further, we assume 2 is a unit of R and R/m is not the field of 3 elements.

LEMMA A. SL(V) is transitive on lines in V .

Proof. Let $L_1 = Rx$ and $L_2 = Ry$ be lines in V . Suppose $V = Rx \oplus R\bar{x}$, $V = Ry \oplus R\bar{y}$ and $y = \alpha x + \beta\bar{x}$. Since y is unimodular either α or β is a unit. Define $\sigma : V \to V$ by either

$$\sigma x = \alpha x + \beta\bar{x}$$
$$\sigma\bar{x} = \alpha^{-1}\bar{x}$$

or

$$\sigma x = \alpha x + \beta\bar{x}$$
$$\sigma x = -\beta^{-1}\bar{x} \quad .$$

This completes the proof.

Relative to a basis, say $\{b_1, b_2\}$, of V , the matrices

$$E_{12}(\lambda) = \begin{bmatrix} 1 & \lambda \\ 0 & 1 \end{bmatrix} \quad \text{and} \quad E_{21}(\lambda) = \begin{bmatrix} 1 & 0 \\ \lambda & 1 \end{bmatrix}$$

are called (linear) elementary transvections.

LEMMA B. Any (linear) transvection in $SC(V,A)$ is conjugate to an elementary transvection.

Proof. Pick a basis $\{b_1,b_2\}$ of V. Let τ be a transvection with line L. By Lemma A, there is a σ in $SL(V)$ with $\sigma L = Rb_1$. Hence $\sigma\tau\sigma^{-1}$ is a transvection with line Rb_1 and is elementary relative to $\{b_1,b_2\}$.

COROLLARY. The order, $O(\tau)$, of a transvection τ is a principal ideal.

Recall $SC(V,A)$ is the subgroup generated by transvections τ with $O(\tau) \subset A$. Further $SC(V,A)$ is a normal subgroup of $SL(V)$.

THEOREM C. Let $\{\tau_\lambda\}_{\lambda\in\Lambda}$ be a family of transvections of order $A_\lambda \subset R$. Let $A = \sum_\lambda A_\lambda$. Then the $SL(V)$-normal subgroup generated by $\{\tau_\lambda\}$ is $SC(V,A)$.

Proof. Consider a single linear transvection τ. Let G be the normal subgroup generated by τ and assume $\tau = E_{12}(\lambda)$ relative to a suitable basis. Clearly $G \le SC(V,R\lambda)$. For ε a unit in R, form $\Theta\tau\Theta^{-1}$ where

$$\Theta = \begin{bmatrix} \varepsilon & 0 \\ 0 & 1 \end{bmatrix}.$$

Then $E_{12}(\varepsilon\lambda)$ is in G. Since, if α is in m then $-1+\alpha$ is a unit, it is easy to see $E_{12}(\beta\lambda)$ is in G for all β in R. Hence $SC(V,R\lambda) \leq G$.

Let G be generated by $\{\tau_\lambda\}$. As above, $G \leq SC(V,A)$. We claim that if τ is in $SC(V,A)$ then τ is in G. Let τ be a transvection with $O(\tau) \subset A$. Assume $\tau = E_{12}(\mu)$ by Lemma B. Since each A_λ above is principal, we have $SC(V,A_\lambda) \leq G$ for each λ. Since

$$A = \sum_\lambda A_\lambda$$

and μ is in A, we have

$$\mu = \sum_{\text{finite}} \mu_\lambda$$

where μ_λ is in A_λ. But $E_{12}(\mu_\lambda)$ is in $SC(V,A_\lambda) \leq G$. Hence

$$\tau = E_{12}(\mu) = \prod_{\text{finite}} E_{12}(\mu_\lambda)$$

is in G.

<u>THEOREM D</u>. Let G be an $SL(V)$-normal subgroup of $GL(V)$ with $O(G) = A$. Then

$$SC(V,A) \leq G \leq GC(V,A) .$$

<u>Proof</u>. (Klingenberg [89]) The proof involves a series of steps and will be given in the context of 2 by 2 matrices.

(a) Suppose G contains an element ρ of the form $\begin{bmatrix} a & u \\ 0 & a+v \end{bmatrix}$ and suppose $A = Rv$. Then

$$SC(V,A) \leq G .$$

Proof. Let $\tau = \begin{bmatrix} 1 & -(a+v) \\ 0 & 1 \end{bmatrix}$. Then $\rho\tau\rho^{-1}\tau^{-1} = \begin{bmatrix} 1 & v \\ 0 & 1 \end{bmatrix}$ is in G. Thus by Theorem C,

$$SC(V,A) \leq G .$$

(b) Suppose $0(G) = A \neq R$ and

$$\sigma = \begin{bmatrix} b & x \\ y & b+z \end{bmatrix}$$

(Note: b is a unit since $A \neq R$.)

is in G. Then G contains $\begin{bmatrix} 1 & y-x \\ 0 & 1 \end{bmatrix}$, $\begin{bmatrix} 1 & y-2x-z \\ 0 & 1 \end{bmatrix}$, and $\begin{bmatrix} 1 & y-2x+z \\ 0 & 1 \end{bmatrix}$.

Remark. If the above is true then if

$$(*) \qquad \begin{aligned} a &= y - x \\ b &= y - 2x - z \\ c &= y - 2x + z , \end{aligned}$$

we have by Theorem C

$$SC(V,Ra) , \quad SC(V,Rb) \text{ and } SC(V,Rc)$$

in G and thus

$$SC(V,A) \le G$$

where $A = Rа + Rb + Rc$. But the matrix of coefficients of (*) is invertible. Hence $A = Rx + Ry + Rz$. This ideal is precisely the order of σ , $O(\sigma)$. Thus

$$SC(V,O(\sigma)) \le G .$$

We now prove statement (b).

Proof. Since σ is in G and G is SL(V)-normal, G contains

$$\begin{bmatrix} 0 & -1 \\ 1 & 0 \end{bmatrix}\begin{bmatrix} b & x \\ y & b+z \end{bmatrix}\begin{bmatrix} 0 & 1 \\ -1 & 0 \end{bmatrix}\begin{bmatrix} b & x \\ y & b+z \end{bmatrix} = \begin{bmatrix} -y^2+b(b+z) & (b+z)(x-y) \\ b(y-x) & b(b+z)-x^2 \end{bmatrix} .$$

Let λ for the moment be undetermined. Set

$$\begin{aligned}
\bar{y} &= b(y - x) \\
\bar{b} &= (b^2 + bz - y^2) - \lambda\bar{y} = b^2 + bz - y^2 - \lambda b(y - x) \\
\bar{z} &= (b^2 + bz - x^2) - \bar{b} + \lambda\bar{y} = y^2 - x^2 + 2\lambda b(y - x) \\
\bar{x} &= (b + z)(x - y) + \lambda^2\bar{y} - \lambda\bar{z} = (x - y)[b(1 + \lambda^2) + z + \lambda(y + x)] .
\end{aligned}$$

Then, the above matrix becomes

$$\bar{\sigma} = \begin{bmatrix} \bar{b}+\lambda\bar{y} & -\lambda^2\bar{y}+\bar{x}+\lambda\bar{z} \\ \bar{y} & \bar{b}-\lambda\bar{y}+\bar{z} \end{bmatrix} .$$

On the other hand,

$$\begin{bmatrix} 1 & \lambda \\ 0 & 1 \end{bmatrix} \begin{bmatrix} \bar{b} & \bar{x} \\ \bar{y} & \bar{b}+\bar{z} \end{bmatrix} \begin{bmatrix} 1 & -\lambda \\ 0 & 1 \end{bmatrix} = \begin{bmatrix} \bar{b}+\lambda\bar{y} & -\lambda^2\bar{y}+\bar{x}+\lambda\bar{z} \\ \bar{y} & \bar{b}-\lambda\bar{y}+\bar{z} \end{bmatrix} = \bar{\sigma} \; .$$

Thus, if $\sum = \begin{bmatrix} \bar{b} & \bar{x} \\ \bar{y} & \bar{b}+\bar{z} \end{bmatrix}$ then

$$\sum = \begin{bmatrix} 1 & \lambda \\ 0 & 1 \end{bmatrix}^{-1} \bar{\sigma} \begin{bmatrix} 1 & \lambda \\ 0 & 1 \end{bmatrix}$$

is in G . If we select $\lambda = -(x + y)/2b$ then

$$\sum = \begin{bmatrix} (b^2+bz)-(y^2+x^2)/2 & (b+z)(x-y)-(x^2-y^2)\frac{(x+y)}{4b} \\ b(y-x) & (b^2+bz)-(y^2+x^2)/2 \end{bmatrix} .$$

That is,

$$\sum = \begin{bmatrix} a & u \\ v & a \end{bmatrix} .$$

Let τ in $SL(V)$ be given by

$$\begin{bmatrix} 1+v & -a(1+v)^{-1}(2+v) \\ 0 & (1+v)^{-1} \end{bmatrix} .$$

(Note $\sum$ is in G where $O(G) = A \neq R$. Thus u and v are in $A \subset m$ and $1 + u$, $1 + v$ are units.)

Then

$$\rho = \tau^{-1}\Sigma^{-1}\tau\Sigma$$

is in G and

$$\rho = (a^2 - uv)^{-1} \begin{bmatrix} a^2-a^2v(2+v)-uv(1+v)^{-2} & * \\ 0 & a^2-a^2v(2+v)-uv(1+v)^2 \end{bmatrix}$$

$$= \begin{bmatrix} \bar{a} & w \\ 0 & \bar{a}+\alpha \end{bmatrix}$$

where $R\alpha = Rv = R(y - x)$. We apply step (a) to ρ and obtain

$$\begin{bmatrix} 1 & y-x \\ 0 & 1 \end{bmatrix}$$

in G .

We have found the first of the three matrices.

Returning to

$$\sigma = \begin{bmatrix} b & x \\ y & b+z \end{bmatrix}$$

we find G also contains

$$\begin{bmatrix} 1 & 0 \\ \pm 1 & 1 \end{bmatrix} \sigma \begin{bmatrix} 1 & 0 \\ \mp 1 & 1 \end{bmatrix} = \begin{bmatrix} b\mp x & x \\ y-x\mp z & b\pm x+z \end{bmatrix} .$$

An argument similar to the above will produce

$$\begin{bmatrix} 1 & y-2x-z \\ 0 & 1 \end{bmatrix} \quad \text{and} \quad \begin{bmatrix} 1 & y-2x+z \\ 0 & 1 \end{bmatrix} .$$

This completes Case (b). By the remark following the statement of (b)

$$SC(V,O(\sigma)) \leq G$$

when σ is in G and $O(\sigma) \neq R$.

(c) Suppose $O(G) = R$. Then $SL(V) \leq G$.

Proof. If $\sigma = \begin{bmatrix} a & b \\ c & d \end{bmatrix}$ is in G then $O(\sigma) = R(a - d) + Rb + Rc$. Since $O(G) = R$, G must contain at least one σ with $O(\sigma) = R$ — otherwise $O(\sigma) \subset m$ for each σ in G and thus $O(G) \subset m$. Hence, letting

$$\sigma = \begin{bmatrix} a & b \\ c & a+x \end{bmatrix}$$

then $Rb + Rc + Rx = R$ and b, c or x is a unit. By suitable conjugation we can produce in G an element of the form

$$\alpha = \begin{bmatrix} 0 & x \\ 1 & y \end{bmatrix} \quad (x \text{ a unit}) .$$

Let

$$\tau = \begin{bmatrix} yd & d^{-1} \\ -d & 0 \end{bmatrix}$$

for d a unit. Then τ is in $SL(V)$ and

$$\rho = \tau^{-1}\alpha^{-1}\tau\alpha = \begin{bmatrix} -x^{-1}d^{-2} & -y-x^{-1}d^{-2}y \\ 0 & -xd^{2} \end{bmatrix}$$

is in G. We want to apply (a). Thus, d in R^* must be selected satisfying

$$x^{-1}d^{-2} - xd^{2} = \text{unit} .$$

Equivalently,

$$x^{2}d^{4} - 1 = \text{unit} .$$

Since 2 is a unit and $R/m \neq$ field of 3 elements there is a d satisfying

$$(xd^{2})^{2} \not\equiv 1 \mod m$$

provided $R/m \neq$ field of 5 elements. (The case $R/m =$ field of 5 elements is handled below.) Thus, with the hypothesis of this section

$$SL(V) \subset G$$

provided $R/m \neq$ field of 5 elements.

Special Case: $R/m =$ field of 5 elements.

If $\Pi_m x^2 \neq 1$ then there is a d satisfying $x^2 d^4 - 1 = \text{unit}$. Thus, suppose $\Pi_m x = \pm 1$. Then select d a unit of R such that $\Pi_m xd^2 = 1$. Besides the element ρ above, G contains $\bar{\rho}$ which arises from d being replaced by $x^{-1}d^{-1}$. Then $\rho\bar{\rho}$ is also in G and has the form

$$\rho\bar{\rho} = \begin{bmatrix} 1 & y(1+x^{-1}d^{-2})^2 \\ 0 & 1 \end{bmatrix} .$$

Hence $\rho\bar{\rho}$ is a transvection of order (y). If $(y) = R$ we have by Theorem C, $SC(V,R) = SL(V) \subset G$. If $(y) \subset m$, $SC(V,O(y)) \subset G$. Then

$$\begin{bmatrix} 0 & x \\ 1 & y \end{bmatrix} \begin{bmatrix} 1 & -y \\ 0 & 1 \end{bmatrix}$$

is also in G. Thus we have an element σ of the form

$$(*) \qquad \begin{bmatrix} 0 & x \\ 1 & 0 \end{bmatrix}$$

with $\Pi_m x = \pm 1$. Then G also contains (see $\rho\bar{\rho}$) $(y = 0)$

$$\Delta = \begin{bmatrix} -x^{-1}d^{-2} & -y-x^{-1}d^{-2}y \\ 0 & -xd^2 \end{bmatrix} .$$

(Note $\Delta = \rho$ of previous page.)

Set $d = 1$. Then

$$A = O(\Delta) = (x^2 - 1) \subset m$$

and

$$SC(V,A) \subset G .$$

We conclude with 2 cases:

(a) $A = (x + 1)$.

(b) $A = (x - 1)$.

In case (a), choose a basis such that (*) becomes

$$\begin{bmatrix} 1 & 2 \\ 1 & -2 \end{bmatrix} \begin{bmatrix} 0 & x \\ 1 & 0 \end{bmatrix} \begin{bmatrix} 1/2 & 1/2 \\ 1/4 & -1/4 \end{bmatrix} = \begin{bmatrix} (4+x)/4 & (4-x)/4 \\ (-4+x)/4 & (-4-x)/4 \end{bmatrix} .$$

Thus, there is an element whose image under λ_A is of the form

$$\begin{bmatrix} 2 & 0 \\ 0 & 2-4 \end{bmatrix} .$$

In case (b), choose a basis such that (*) has the form:

$$\begin{bmatrix} 1 & 1 \\ 1 & -1 \end{bmatrix} \begin{bmatrix} 0 & x \\ 1 & 0 \end{bmatrix} \begin{bmatrix} 1/2 & 1/2 \\ 1/2 & -1/2 \end{bmatrix} = \begin{bmatrix} (1+x)/2 & (1-x)/2 \\ (-1+x)/2 & (-1-x)/2 \end{bmatrix} .$$

Thus under λ_A this has the form

$$\begin{bmatrix} 1 & 0 \\ 0 & 1-2 \end{bmatrix} \quad \text{(where } x = 1 \text{).}$$

In both cases,

$$SC(V/AV,R/A) \subset \lambda_A G$$

and thus

$$SC(V,R) = SL(V) \subset G \ .$$

This completes the proof and this section. A detailed discussion of the normal subgroups of $GL_2(R)$ when 2 is in m or R/m = field of 3 elements is given by Lacroix [96].

V. THE ORTHOGONAL GROUP: NORMAL SUBGROUPS AND AUTOMORPHISMS

(A) INTRODUCTION

The purpose of this chapter is to provide a description of the generators, normal subgroups and automorphisms of the orthogonal group $O(V)$ of a symmetric inner product space having isotropic vectors.

The first several sections examine the generators of $O(V)$, i.e., symmetries, reflections, unitary transvections, etc. For these generators and the commutator subgroups we also examine the situation when no assumption is made concerning the existence of isotropic vectors, i.e., no assumption on hyperbolic rank of V being ≥ 1 .

However, the normal subgroup theory is developed under the constraint that the hyperbolic rank of V is ≥ 1 . We also rule out various choices of the residue class fields (as in (IV)) which have been known to create difficulties. We will find that under the above hypothesis the normal subgroups are sandwiched between congruence subgroups determined by the ideals of the local ring R . The results of this nature for local rings and the orthogonal group were first developed by Klingenberg [90]; however, we follow a more recent treatment by James [87].

The assumptions necessary for the normal subgroups are continued into the sections examining the automorphisms of $O(V)$. The approach we employ is the involution technique of Dieudonné and Rickart which was extended to local rings by Keenan [88].

Under suitable hypothesis on hyperbolic rank, existence of units and dimension of the space V, we find the automorphisms of $O(V)$ are compositions of radial automorphisms and conjugations of R-semi-linear isomorphisms of V which preserve orthogonality.

(B) SYMMETRIC INNER PRODUCT SPACES

We begin this section by recalling some of the results in (III). Throughout this chapter R will denote a local ring with maximal ideal m, residue class field $k = R/m$ and 2 a unit. Further, $V = (V,\beta)$ will denote an R-space of dimension n with a symmetric inner product

$$\beta : V \times V \to R .$$

If e is in V then $\beta(e,e)$ is the norm of e.

Suppose e in V is unimodular. Two important situations occur depending on whether $\beta(e,e) = \text{unit}$ or $\beta(e,e) = 0$. (Note: $\beta(e,e)$ may also be a nonzero non-unit.)

(1) If $\beta(e,e) = v$ a unit of R, then e is called nonisotropic and Re splits off V as an orthogonal summand, i.e.,

$$V = Re \perp (Re)^{\perp} \simeq \langle v \rangle \perp (Re)^{\perp} .$$

If $e = e_1$, we can find additional unimodular vectors $e_2,\ldots,e_n$ with $\beta(e_i,e_i) = v_i$ where v_i is a unit for $2 \leq i \leq n$ such that

$$\begin{aligned} V &= Re_1 \perp Re_2 \perp \cdots \perp Re_n \\ &\simeq \langle v_1 \rangle \perp \langle v_2 \rangle \perp \cdots \perp \langle v_n \rangle \end{aligned}$$

and V is the orthogonal sum of symmetric inner product spaces of dimension one.

(2) If $\beta(e,e) = 0$ then e is called _isotropic_ and we can find a unimodular f in V with $\beta(e,f) = 1$. The plane $H = Re \oplus Rf$ is a hyperbolic plane and V splits as

$$V = H \perp W .$$

The plane $H \simeq \langle 1 \rangle \perp \langle -1 \rangle$. If

$$V = H_1 \perp H_2 \perp \cdots \perp H_t \perp W$$

where $H_1,\ldots,H_t$ are hyperbolic planes then V is said to have _hyperbolic rank_ $\geq t$. If $V = H_1 \perp \cdots \perp H_m$ ($2m = n$) is the orthogonal sum of hyperbolic planes then V is said to be _hyperbolic_ or _split_.

Every symmetric inner product space V has a decomposition

$$V = W \perp Z$$

where W is a symmetric inner product space having no unimodular isotropic vectors and Z is a split space. Further, by Witt cancellation (III.19) (see remark after (III.19) also) W and Z are uniquely determined up to isometry.

If (V,β) is a symmetric inner product space then

$$O(V) = \{\sigma \text{ in } GL(V) \mid \beta(\sigma x,\sigma y) = \beta(x,y) \text{ for all } x,y \text{ in } V\}$$

is the <u>orthogonal group</u> of V. An element σ satisfying

(a) σ in $GL(V)$

(b) $\beta(\sigma x,\sigma y) = \beta(x,y)$ for all x and y in V,

i.e., σ is in $O(V)$, is an <u>isometry</u> of V.

If A is an ideal of R the ring morphism $\Pi : R \to R/A$ induces an R-morhpism $\Pi_A : V \to V/AV$ of symmetric inner product spaces where the inner product $\bar{\beta} : V/AV \times V/AV \to R/A$ is given by

$$\bar{\beta}(\Pi_A x,\Pi_A y) = \Pi_A\beta(x,y) .$$

In turn, Π_A induces a group morphism

$$\lambda_A : O(V) \to O(V/AV) .$$

If $\sigma : V \to V$ is an isometry and V splits as $V = U \perp W$ then σ is uniquely determined by its restrictions

$$\rho = \sigma|_U : U \to V ,$$
$$\tau = \sigma|_W : W \to V ,$$

and σ is written as $\sigma = \rho \perp \tau$.

If $\{b_1,\ldots,b_n\}$ is a basis for (V,β) then

$$B = [\beta_{ij}] = [\beta(b_i,b_j)]$$

is the matrix representative of β relative to $\{b_1,\ldots,b_n\}$. If σ in $GL(V)$ has matrix $\sum$ relative to $\{b_1,\ldots,b_n\}$ then σ is in $O(V)$ if and only if

$$\sum B \sum^t = B .$$

Since β is an inner product, $\det(B)$ is a unit. Thus, from $\sum B \sum^t = B$, we conclude $\det(\sum)$, and hence $\det(\sigma)$, is ± 1 . Occasionally, (V,β) is denoted by $(V,\langle B\rangle)$ or simply $\langle B\rangle$ (this latter notation is usually employed when $\dim(V) = 1$).

The isometries in $O(V)$ having determinant 1 form a subgroup $SO(V)$ called the <u>special orthogonal group</u> of V or the <u>group of rotations</u>. If σ in $O(V)$ has $\det(\sigma) = 1$ then σ is called a <u>rotation</u> and if $\det(\sigma) = -1$ then σ is called a <u>reflection</u>. Observe $[O(V) : SO(V)] = 2$.

(C) SYMMETRIES AND THE ORTHOGONAL GROUP

In this section we describe a set of generators of $O(V)$. However, we make no assumption on the hyperbolic rank of V as in (III.D).

Recall, from (II.E), an element σ in $GL(V)$ is called an involution if $\sigma^2 = I$. If σ is an involution then

$$P(\sigma) = \{x \text{ in } V \mid \sigma x = x\}$$

is the positive space of σ ,

$$N(\sigma) = \{x \text{ in } V \mid \sigma x = -x\}$$

is the negative space of σ and $V = P(\sigma) \oplus N(\sigma)$. If, in addition, σ is in $O(V)$ then it is easy to check that

$$V = P(\sigma) \perp N(\sigma)$$

and $\sigma = i_{P(\sigma)} \perp -i_{N(\sigma)}$. Conversely, if V splits as $V = U \perp W$ then $\sigma = i_U \perp (-i_V)$ is an involution in $O(V)$.

The most important involutions in $O(V)$ are the symmetries or hyperplane reflections which we now define. Suppose x in V has $\beta(x,x)$ a unit, i.e., x is nonisotropic. Define

$$\sigma_x : V \to V$$

by

$$\sigma_x(z) = z - 2\,\frac{\beta(x,z)}{\beta(x,x)}\,x \qquad \text{for } z \text{ in } V .$$

Then σ_x is a <u>symmetry</u> with $P(\sigma_x) = (Rx)^\perp$ and $N(\sigma_x) = Rx$. (Note: σ_x is an extremal involution in the sense of (II.E).) Pictorially, σ_x induces a reflection about the hyperplane $(Rx)^\perp$:

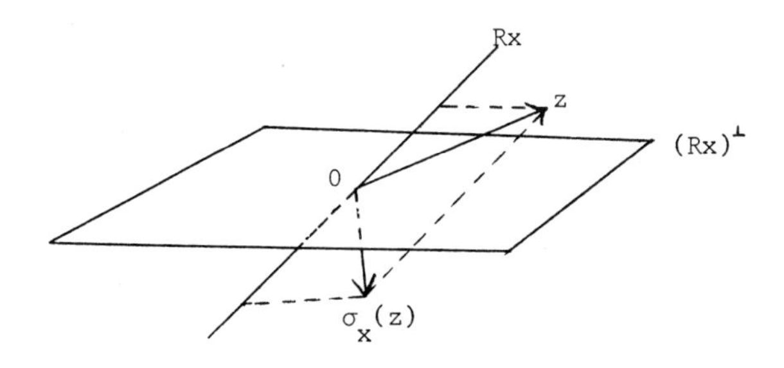

The symmetry σ_x leaves every vector in the hyperplane $(Rx)^\perp$ fixed and reverses each vector in the line Rx . Observe

$$\sigma_x = -i_{Rx} \perp i_{(Rx)^\perp} .$$

<u>(V.1)</u> <u>THEOREM</u>. (On Symmetries) Let x and y be nonisotropic vectors in V . Then

(a) The determinant of σ_x is -1 . Hence σ_x is a reflection.

(b) If τ is in $O(V)$ then

$$\tau\sigma_x\tau^{-1} = \sigma_{\tau x} .$$

(c) Since σ_x is an involution, $\sigma_x^{-1} = \sigma_x$. Further, $\sigma_{\lambda x} = \sigma_x$ for λ a unit of R .

(d) For z in V ,

$$\sigma_x\sigma_y(z) = z - 2\,\frac{\beta(x,z)}{\beta(x,x)}\,x - 2\,\frac{\beta(y,z)}{\beta(y,y)}\,y + 4\,\frac{\beta(y,z)\beta(x,y)}{\beta(y,y)\beta(x,x)}\,x$$

gives the product of two symmetries. In particular, $\sigma_x\sigma_y = \sigma_y\sigma_x$ if and only if $Rx = Ry$ or $\beta(x,y) = 0$.[1]

(e) Suppose $\beta(x,x) = \beta(y,y)$ and $x + y$ is nonisotropic. Then $\sigma_{x+y}(x) = -y$ and $\sigma_{x+y}(y) = -x$. In particular, if $\beta(x,y) = 0$ then $\beta(x + y,x + y) = 2\beta(x,x)$, $x + y$ is nonisotropic and $\sigma_y\sigma_{x+y}(x) = y$.

Proof. The above are straightforward.

(V.2) COROLLARY. Suppose x and y are in V and

$$\beta(x,x) = \beta(y,y) = \text{unit} .$$

Then there is a product of two symmetries carrying x to y .

Proof. First express x as a sum of two orthogonal vectors

$$e = \frac{1}{2}(x + y) , \quad f = \frac{1}{2}(x - y) .$$

Then $\beta(x,x) = \beta(e,e) + \beta(f,f)$. Since $\beta(x,x)$ is a unit and R is

[1] The converse follows from the discussion in (II.E).

local, either $\beta(e,e)$ or $\beta(f,f)$ is a unit. Assume $\beta(e,e)$ is a unit. The other case is handled similarly. Then

$$\sigma_e(x) = \sigma_e(e + f) = (e + f) - 2\,\frac{\beta(e,e+f)}{\beta(e,e)}\,e = -e + f = -y\ ,$$

since $\beta(e,e + f) = \beta(e,e) + \beta(e,f) = \beta(e,e)$. Then $\sigma_y\sigma_e x = \sigma_y(-y) = y$.

Since symmetries are in $O(V)$, the above corollary may be restated as the following theorem on the action of $O(V)$ on unimodular vectors of <u>unit</u> norm.

<u>(V.3)</u> <u>THEOREM</u>. The orthogonal group $O(V)$ acts as a transformation group on the nonisotropic vectors of V . Two nonisotropic vectors x and y are in the same $O(V)$-orbit if and only if $\beta(x,x) = \beta(y,y)$.

Observe, the above result does not require the hypothesis that the hyperbolic rank of V is ≥ 1 , which was necessary in (III.17). However, (V.3) applies only to the nonisotropic vectors of V , i.e., vectors of unit norm, and not, as does (III.17), to the unimodular vectors.

We now consider the generation of $O(V)$ by symmetries.

<u>(V.4)</u> <u>PROPOSITION</u>. Let $\dim(V) = 2$.

(a) Each element τ of $O(V)$ may be written as a product of one or two symmetries as follows:

(i) If $\det(\tau) = -1$, then τ is a symmetry.

(ii) If $\det(\tau) = 1$, i.e., τ is in SO(V) , then $\tau = \sigma_x\sigma$ where x is a fixed (but arbitrary) nonisotropic vector and σ , σ_x are symmetries.

(b) SO(V) is an Abelian group.

Proof. (a) Let x be a fixed nonisotropic vector and $V = Rx \perp Ry$. Let α be in SO(V) , i.e., $\det(\alpha) = 1$. Then $\beta(\alpha^{-1}x,\alpha^{-1}x) = \beta(x,x) =$ unit . As in the proof of (V.2), there is a symmetry σ_e with $\sigma_e(\alpha^{-1}x) = -x$.

Consider $\sigma_e(\alpha^{-1}y) = ax + by$. Since $\beta(\alpha^{-1}x,\alpha^{-1}y) = \beta(x,y) = 0$, one has $a = 0$. Since $\beta(\alpha^{-1}y,\alpha^{-1}y) = \beta(y,y) =$ unit , one has $b = \pm 1$. But $\det(\alpha) = 1$, hence $\det(\alpha^{-1}) = 1$ and consequently $b = 1$ (since $\det(\sigma_e\alpha^{-1}) = \det(\sigma_e)\det(\alpha^{-1}) = (-1)(1) = -1$). Thus

$$\sigma_e\alpha^{-1}(x) = -x$$
$$\sigma_e\alpha^{-1}(y) = y .$$

Therefore $\sigma_x\sigma_e\alpha^{-1} = I$ and $\alpha = \sigma_x\sigma_e$.

Suppose $\bar{\alpha}$ is in O(V) and not in SO(V) . Hence $\det(\bar{\alpha}) = -1$. Then $\sigma_x\bar{\alpha}$ is in SO(V) and, by the above,

$$\sigma_x\bar{\alpha} = \sigma_x\sigma_e .$$

Equivalently,

$$\bar{\alpha} = \sigma_e .$$

To show (b), let τ_1 and τ_2 be in $SO(V)$ and, by the proof of (a), write these isometries as

$$\tau_1 = \sigma\sigma_1 = \sigma_2\sigma_3 , \quad \tau_2 = \sigma\sigma_2$$

where σ , σ_1 , σ_2 and σ_3 are symmetries. Then

$$\begin{aligned} \tau_2\tau_1\tau_2^{-1} &= (\sigma\sigma_2)(\sigma_2\sigma_3)(\sigma_2\sigma) \\ &= (\sigma\sigma_3)(\sigma_2\sigma) \\ &= (\sigma\sigma_1)(\sigma\sigma) \\ &= \tau_1 \end{aligned}$$

<u>(V.5)</u> <u>THEOREM</u>. (Klingenberg) Let $\dim(V) = n$. Each isometry τ in $O(V)$ may be written as a product of $\leq 2n - 2$ symmetries. Hence, the symmetries generate $O(V)$. Further, τ is in $SO(V)$ if and only if the number of symmetries in a product giving τ is even.

<u>Proof</u>. The proof is an induction on $\dim(V)$. The case $\dim(V) = 2$ is given in (V.4).

Let x be a nonisotropic vector in V . Let τ be in $O(V)$. Then, as observed previously, $\tau x - x$ or $\tau x + x$ is nonisotropic. Thus, for $\varepsilon = 1$ or $\varepsilon = -1$, assume

$$\tau x + \varepsilon x$$

is nonisotropic. Let $\sigma = \sigma_{\tau x + \varepsilon x}$. Then, a direct computation gives

$$\sigma\tau x = -\varepsilon x \ .$$

Let $H = (Rx)^{\perp}$. Since $\sigma\tau : Rx \to Rx$, $\sigma\tau$ will carry the orthogonal complement H of Rx to H . By induction

$$\sigma\tau|_H = \sigma_1\sigma_2\ldots\sigma_t$$

where $t \leq 2(n - 1) - 2 = 2n - 4$ and the σ_i are symmetries in $O(H)$. Extend each σ_i to an element of $O(V)$ by $\sigma_i(x) = x$.

Then, if $\varepsilon = -1$

$$\tau = \sigma\sigma_1\sigma_2\ldots\sigma_t$$

and, if $\varepsilon = 1$

$$\tau = \sigma\sigma_x\sigma_1\sigma_2\ldots\sigma_t \ .$$

In both cases τ is a product of symmetries and the number of factors $\leq 2n - 2$.

If R is a field, then the Cartan-Dieudonné Theorem (for example, see [18], page 27) states that each isometry τ in $O(V)$ may be written as a product of $\leq \dim(V)$ symmetries. It is likely the same result is true for local rings having 2 a unit. Observe the above proof (slightly modified) would give this bound if we could escape using the σ_x in the last step of the proof.

Also observe the above proof and (V.4) indicate that if τ in $O(V)$ is written as

$$\tau = \sigma_1 \sigma_2 \cdots \sigma_t$$

a product of symmetries σ_i then σ_1 may be arbitrarily chosen.

Since each symmetry $\bar{\sigma}$ in $O(V/AV)$ (A an ideal of R) has a pre-image in $O(V)$, i.e., there is a symmetry σ with $\lambda_A \sigma = \bar{\sigma}$, we have the following corollary. Note also $\det(\sigma) = \det(\bar{\sigma})$.

<u>(V.6)</u> <u>COROLLARY</u>. Let A be an ideal of R .

(a) The group morphism

$$\lambda_A : O(V) \to O(V/AV)$$

is surjective.

(b) The restriction

$$\lambda_A \big|_{SO(V)} : SO(V) \to SO(V/AV)$$

is surjective.

We contrast (V.6) with (III.21). Corollary (III.21) required the hyperbolic rank of $V \geq 1$.

(V.7) COROLLARY. Let A be a proper ideal of R. Let τ be in $O(V)$ and suppose

$$\lambda_A \tau = 1 .$$

Then τ is a product of $\leq n - 1$ elements of the form $\sigma_1\sigma_2$, where σ_1 and σ_2 are symmetries and $\lambda_A(\sigma_1\sigma_2) = 1$.

Proof. Let x be a nonisotropic vector of V. Then $\Pi_A(\tau x + x) = (\lambda_A\tau)\Pi_A x + \Pi_A x = 2\Pi_A x$ and $\tau x + x$ is nonisotropic. Set

$$\sigma = \sigma_x \qquad \text{and} \qquad \bar{\sigma} = \sigma_{x+\tau x} .$$

Then, as in the first part of the proof of (V.5),

$$\sigma\bar{\sigma}(\tau x) = x .$$

Thus,

$$\tau = \bar{\sigma}\sigma\bar{\tau}$$

where $\bar{\tau}|_{Rx} = I$. Since $V = Rx \perp (Rx)^{\perp}$ the proof follows by induction.

We conclude this section with a few remarks on the relationship between symmetries and the isometries discussed in (III.D).

Suppose V has a unimodular isotropic vector u and

$$V = H \perp W$$

where $H = Ru \oplus Rv$, $\beta(u,v) = 1$ and $\beta(u,u) = \beta(v,v) = 0$.

Suppose x is <u>nonisotropic</u> and $\beta(x,u) = 0$. Using the formula for the product of two symmetries given in (V.1)(d), the unitary transvection $\sigma_{u,x}$ (from (III.D)) may be expressed as a product of two symmetries. Precisely,

$$\sigma_{u,x} = \sigma_x \sigma_{-x+\frac{1}{2}\beta(x,x)u}$$

or

$$\sigma_{u,x} = \sigma_{-x-\frac{1}{2}\beta(x,x)u} \sigma_x \ .$$

If α is a unit in R , then the vector $x = u - \alpha v$ is nonisotropic. Thus, we can form the symmetry σ_x . Observe

$$\sigma_x(u) = \alpha v$$
$$\sigma_x(v) = \alpha^{-1} u$$

and

$$\sigma_x(w) = w \quad \text{for all } w \text{ in } W \ .$$

By selecting α we may construct the isometries Δ and Φ_ε in (III.D). Precisely,

$$\Delta = \sigma_{u-v} \qquad (\alpha = 1)$$

and if $\alpha = \varepsilon$

$$\Phi_\varepsilon = \sigma_{u-v}\sigma_{u-\varepsilon v} = \sigma_{u-\varepsilon v}\sigma_{u-v} \ .$$

Hence Δ is a symmetry and Φ_ε is a rotation.

Observe the above is a special case of (V.4).

Conversely, if x is nonisotropic in W (hence both $\beta(u,x) = 0$ and $\beta(v,x) = 0$), then the symmetry σ_x is given by $\sigma_x = \Delta^{-1}\Phi_{-\frac{1}{2}\beta(x,x)}\sigma_{v,-x}\sigma_{u,-\frac{2x}{\beta(x,x)}}\sigma_{v,-x}$. This is tedious to check but follows in a straightforward manner by computing the action of both sides of the equation on u,v and then on an arbitrary w in W .

(D) THE CLIFFORD ALGEBRA AND ISOMETRIES

The purposes of this section are initially the construction of the Clifford algebra associated with a symmetric inner product space (V,β) and finally a discussion of how various isometries arise naturally as inner-automorphisms of the Clifford algebra. We are indebted to D. G. James for pointing out the connection between the Clifford algebra of (V,β) and isometries in $O(V)$.

We let R be a local ring and (V,β) a symmetric inner product space.

A (<u>free</u>) <u>algebra</u> A over R is an R-space A which is provided with a ring structure satisfying the following:

(a) $\alpha(xy) = (\alpha x)y = x(\alpha y)$,

(b) $(\alpha 1_A)x = \alpha x = x(\alpha 1_A)$,

(c) 1_A is R-unimodular

for x and y in A , α in R and 1_A the identity of A . We assume 1_A is unimodular as a vector over R , hence $\alpha 1_A = \beta 1_A$ implies $\alpha = \beta$ for α and β in R . We write $\alpha 1_A$ as α for α in R .

A mapping $\sigma : A \to B$ from an R-algebra A to an R-algebra B is an <u>algebra morphism</u> if

(a) σ is an R-morphism,

(b) σ is a ring morphism (we assume $\sigma(1_A) = 1_B$).

We say an algebra A is <u>compatible</u> with the symmetric inner product space (V,β) provided

(a) V is an R-subspace of A (via a suitable identification),

(b) for each x in V the ring structure on A satisfies

$$x^2 = \beta(x,x) .$$

Suppose an algebra A is compatible with (V,β) . Then, for x and y in V

$$\begin{aligned}(x + y)^2 &= \beta(x + y,x + y)\\ &= \beta(x,x) + \beta(y,y) + 2\beta(x,y)\end{aligned}$$

$$= x^2 + y^2 + 2\beta(x,y) .$$

Thus, for x and y in V

$$xy + yx = 2\beta(x,y) .$$

In particular, if x is orthogonal to y then

$$xy = -yx .$$

(V.8) LEMMA. Let A be an algebra compatible with (V,β) . Then

(a) For x and y in V , $xy + yx = 2\beta(x,y)$. Further, x is orthogonal to y if and only if $xy = -yx$ in A .

(b) x in V is nonisotropic if and only if x is a unit in A . Further, if x is nonisotropic then $x^{-1} = \beta(x,x)^{-1}x$.

(c) If $\{x_1,\ldots,x_n\}$ is an orthogonal basis of V then, in A ,

$$x_i x_j = \begin{cases} -x_j x_i & \text{if } i \neq j \\ \beta(x_i,x_i) & \text{if } i = j . \end{cases}$$

Proof of (b). If x in V is nonisotropic then $x^2 = xx = \beta(x,x) = \text{unit}$. Thus $x(x\beta(x,x)^{-1}) = 1$. Conversely, suppose $xy = 1$ and $\beta(x,x)$ is not a unit. Then $\beta(x,x)$ is in the maximal ideal m of R . Then $1 = xy = x(xy)y = x^2y^2 = \beta(x,x)y^2$. Since y^2 is in A , we have 1 in mA . Then $A = mA$ and by Nakayama's Lemma (I.1), $A = 0$. This is a contradiction since V is in A . Thus $\beta(x,x)$ is a unit.

An R-algebra C which is compatible with (V,β) is called a <u>Clifford algebra for</u> (V,β) if C satisfies the following mapping property: If A is any algebra compatible with (V,β) then there exists a <u>unique</u> R-algebra morphism

$$\phi : C \to A$$

with $\phi(x) = x$ for all x in V .

If Clifford algebras C and $\bar{C}$ exist it is easy to see there is an R-algebra isomorphism $\psi : C \to \bar{C}$ with $\psi|_V$ = identity .

<u>(V.9) THEOREM</u>. Let (V,β) be a symmetric inner product space over R . Then (V,β) has a Clifford algebra.

<u>Proof</u>. We give the construction of O'Meara ([22], pp. 133-35).[1]

Let (V,β) have an orthogonal basis $\{x_1,\ldots,x_n\}$.

We will construct an algebra $C(V)$, which will be our Clifford algebra. This algebra will have dimension 2^n as an R-space. We begin by introducing an index set for the basis of this algebra.

Let $\Pi = \{0,1\}$ be the finite field of 2 elements. Let

$$\Phi = \Pi \times \Pi \times \cdots \times \Pi \quad \text{(n factors)}$$

[1] See Lam ([18], Chapter V) for a construction of $C(V)$ via the tensor algebra.

be the Cartesian product. Note $|\Phi| = 2^n$ and Φ is an Abelian group under coordinate-wise addition. If α is in Π and δ is a unit of R, define

$$\delta^{\alpha} = \begin{cases} \delta & \text{if} \quad \alpha = 1 \\ 1 & \text{if} \quad \alpha = 0 . \end{cases}$$

Let $C(V)$ be an R-space of dimension 2^n. Select a basis for $C(V)$ and index this basis by the elements of Φ. It is well-known that given any R-space A, then A may be made into an algebra (possibly without identity) by selecting, for a fixed basis, say $\{y_1,\dots,y_m\}$, m^3 elements α_{ijk} in R, requiring

$$y_i y_j = \textstyle\sum_{k=1}^{m} \alpha_{ijk} y_k$$

and

$$y_i(y_j y_k) = (y_i y_j) y_k$$

and extending these products linearly to all of A. We now do this for $C(V)$.

From above $C(V)$ has a basis $\{b_\alpha\}_{\alpha\in\Phi}$. Recall V had an orthogonal basis $\{x_1,\dots,x_n\}$. Let $q_i = \beta(x_i,x_i)$. Let $\alpha = \langle\alpha_1,\dots,\alpha_n\rangle$ and $\delta = \langle\delta_1,\dots,\delta_n\rangle$ be in Φ. Define

$$b_\alpha b_\delta = \Big(\prod_{1\le j<i\le n} (-1)^{\alpha_i\delta_j}\Big)\Big(\prod_{1\le i\le n} q_i^{\alpha_i\delta_i}\Big) b_{\alpha+\delta} .$$

It is straightforward to show

$$(b_\alpha b_\delta)b_\rho = b_\alpha(b_\delta b_\rho)$$

and if $0 = \langle 0,0,\ldots,0\rangle$ in Φ then

$$b_\alpha b_0 = b_\alpha = b_0 b_\alpha \ .$$

Thus $C(V)$ is an algebra with identity b_0 .

Let

$$\bar{x}_1 = b_{\langle 1,0,0,\ldots,0\rangle}$$
$$\bar{x}_2 = b_{\langle 0,1,0,\ldots,0\rangle}$$
$$\vdots \qquad \vdots$$
$$\bar{x}_n = b_{\langle 0,0,0,\ldots,1\rangle} \ .$$

Observe the 2^n vectors $\bar{x}_1^{\varepsilon_1}\bar{x}_2^{\varepsilon_2}\cdots\bar{x}_n^{\varepsilon_n}$ where $\varepsilon_i = 0$ or 1 are R-free and span $C(V)$. Further,

$$b_{\langle\varepsilon_1,\varepsilon_2,\ldots,\varepsilon_n\rangle} = \bar{x}_1^{\varepsilon_1}\bar{x}_2^{\varepsilon_2}\cdots\bar{x}_n^{\varepsilon_n} \ .$$

Identify V as a subspace of $C(V)$ by $x_i \to \bar{x}_i$ for $1 \le i \le n$. It is easy to see under this identification $C(V)$ is compatible with V . It remains to show $C(V)$ possesses the mapping property necessary to be a Clifford algebra.

Suppose A is any algebra compatible with (V,β) . The subring A_V of A generated by V is a subalgebra of A containing the identity of A and is composed of all R-linear combinations of

$$x_1^{\varepsilon_1} x_2^{\varepsilon_2} \cdots x_n^{\varepsilon_n}$$

for $\varepsilon_i = 0,1$ and $\{x_1,\ldots,x_n\}$ the basis of V .

Define an R-algebra morphism $\phi : C(V) \to A$ by defining

$$\phi : \bar{x}_1^{\varepsilon_1} \cdots \bar{x}_n^{\varepsilon_n} \to x_1^{\varepsilon_1} \cdots x_n^{\varepsilon_n}$$

and extending linearly. Recall an algebra morphism is uniquely determined by its action on a basis of the algebra. One checks that ϕ is the identity on V via the identification.

<u>(V.10)</u> <u>COROLLARY</u>. Let (V,β) be a symmetric inner product space with orthogonal basis $\{x_1,x_2,\ldots,x_n\}$. Then a Clifford algebra for (V,β) has dimension 2^n and basis

$$\{x_1^{\varepsilon_1} x_2^{\varepsilon_2} \cdots x_n^{\varepsilon_n} \mid \varepsilon_i = 0,1\} .$$

Since, as noted above, a Clifford algebra for (V,β) is unique up to R-algebra isomorphism, we will refer to the algebra C(V) constructed in the proof of (V.9) as <u>the</u> Clifford algebra of (V,β) .

An algebra A is <u>2-graded</u> if, as an R-space, $A = A_0 \oplus A_1$ and

$$A_i A_j \subset A_{i+j}$$

where the subscripts are taken modulo 2 .

The Clifford algebra is 2-graded in the following fashion: Each element x in $C(V)$ is a finite sum of products of elements of V . If each summand in x is a product of an even number of elements of V then x is called <u>even</u>. Let C_0 be the set of even elements of $C(V)$. C_0 is the set of finite sums of all even products of the basis elements $x_1,\ldots,x_n$ of V . Thus C_0 is spanned by

$$\{x_1^{\varepsilon_1} x_2^{\varepsilon_2} \cdots x_n^{\varepsilon_n} \mid \varepsilon_i = 0,1 \text{ and } \varepsilon_1 + \cdots + \varepsilon_n \equiv 0 \mod 2\} .$$

Hence $\dim(C_0) = 2^{n-1}$. The set C_0 is a subalgebra and is called the <u>even Clifford algebra</u> of (V,β) . The <u>odd</u> summand C_1 of $C(V)$ is defined in a similar fashion and $C(V) = C_0 \oplus C_1$ where $C_i C_j \subset C_{i+j}$ (subscripts modulo 2). (Note $\dim(C_1) = 2^{n-1}$ but C_1 is not an algebra.)

If w in $C(V) = C_0 \oplus C_1$ is in C_i then w is called <u>homogeneous of degree</u> i , denoted $\partial w = i$, for $i = 0,1$.

We want to examine automorphisms of $C(V)$ and observe how isometries of V arise from inner-automorphisms of $C(V)$.

The 2-graded algebra $C(V) = C_0 \oplus C_1$ has the <u>basic automorphism</u>

$$\rho : w = w_0 + w_1 \to \bar{w} = w_0 - w_1$$

for w in $C(V)$ where w_0 is in C_0 and w_1 is in C_1. Observe the action of any automorphism Φ of $C(V)$ on $w = w_0 + w_1$ is determined by the action of Φ on the homogeneous parts w_0 and w_1 of w. Thus, we need only describe Φ on homogeneous elements of $C(V)$. Let hC denote the homogeneous elements of $C(V)$. Thus, the basic automorphism is defined by

$$\rho(w) = (-1)^{\partial w} w \quad \text{for } w \text{ in } hC .$$

Suppose x is a homogeneous unit of $C(V)$. Define the <u>inner-automorphism</u> $I_x : C(V) \to C(V)$ by

$$I_x(w) = (-1)^{\partial x \partial w} x w x^{-1} \quad \text{for } w \text{ in } hC .$$

(Note: Bass([40], p.162 and 166) shows that every R-algebra automorphism of $C(V)$ is an inner-automorphism.)

Suppose x is nonisotropic in V. Then, by (V.8), x is invertible in $C(V)$. Also x is homogeneous of degree $\partial x = 1$. Let us examine the action of the inner-automorphism I_x on elements of V. For y in V,

$$\begin{aligned} I_x(y) &= (-1)xyx^{-1} \\ &= (-1)xyx\beta(x,x)^{-1} && \text{by (V.8)} \\ &= -\beta(x,x)^{-1}[2\beta(x,y)x - yxx] && \text{by (V.8)} \\ &= -\beta(x,x)^{-1}[2\beta(x,y)x - \beta(x,x)y] \\ &= y - 2\,\frac{\beta(x,y)}{\beta(x,x)}\,x \\ &= \sigma_x(y) \end{aligned}$$

where σ_x is the symmetry about the hyperplane $(Rx)^{\perp}$. Hence, a symmetry σ_x of V is induced by an inner-automorphism I_x of C(V).

Further, if x_1 and x_2 are nonisotropic vectors in V, then for y in V

(a) $I_{x_1}(y) = I_{-x_1}(y)$

(b) $I_{x_1} I_{x_2}(y) = \sigma_{x_1}\sigma_{x_2}(y)$

(c) $I_{x_1} I_{x_2} = I_{x_1 x_2}$.

Suppose V splits as $H \perp W = V$ where $H = Ru \oplus Rv$ is a hyperbolic plane with $\beta(u,u) = \beta(v,v) = 0$ and $\beta(u,v) = 1$.

Then, by the remarks at the end of (V.C),

$$\Delta = I_{u-v}\Big|_V$$

and

$$\Phi_\varepsilon = I_{u-v} I_{u-\varepsilon v}\Big|_V \qquad (\varepsilon \text{ a unit}).$$

Further, in C(V), $(u - v)(u - \varepsilon v) = -vu - \varepsilon uv$. Applying (c) and (a) above

$$\Phi_\varepsilon = I_{\varepsilon uv+vu}\Big|_V .$$

We examine next the unitary transvections $\sigma_{u,x}$ and $\sigma_{v,x}$. In (V.C) we required that x be nonisotropic in addition to $\beta(x,u) = 0$

($\beta(x,v) = 0$, respectively). We now drop this former condition. We assume V has, of course, the above splitting $V = H \perp W$.

Let x be in V with $\beta(x,u) = 0$. Then in $C(V)$

$$
\begin{aligned}
(1 + \tfrac{1}{2}ux)(1 + \tfrac{1}{2}xu) &= 1 + \tfrac{1}{2}(ux + xu) + \tfrac{1}{4}uxxu \\
&= 1 + \beta(u,x) + \tfrac{1}{4}\beta(x,x)\beta(u,u) \\
&= 1 .
\end{aligned}
$$

Hence, $1 + \frac{1}{2}ux$ is invertible in $C(V)$ with inverse $1 + \frac{1}{2}xu$ (equivalently, $1 - \frac{1}{2}ux$ since $\beta(u,x) = 0$) and homogeneous of degree 0 .

Consider the inner-automorphism $I_{1+\frac{1}{2}ux}$ of $C(V)$ restricted to V :

$$
\begin{aligned}
I_{1+\frac{1}{2}ux}(z) &= (1 + \tfrac{1}{2}ux)z(1 + \tfrac{1}{2}ux)^{-1} \\
&= (1 + \tfrac{1}{2}ux)z(1 + \tfrac{1}{2}xu) \\
&= z + \tfrac{1}{2}uxz + \tfrac{1}{2}zxu + \tfrac{1}{4}uxzxu \\
&= z + \tfrac{1}{2}(uxz + uzx) + \tfrac{1}{2}(zxu - uzx) \\
&\quad + \tfrac{1}{4}ux(2\beta(x,z) - xz)u \\
&= z + \beta(x,z)u - \beta(u,z)x + \tfrac{1}{2}\beta(x,z)uxu - \tfrac{1}{4}ux^2zu \\
&= z + \beta(x,z)u - \beta(u,z)x - \tfrac{1}{4}\beta(x,x)uzu \\
&= z + \beta(x,z)u - \beta(u,z)x - \tfrac{1}{2}\beta(x,x)\beta(u,z)u \\
&= \sigma_{u,x}(z)
\end{aligned}
$$

A similar expression can be obtained for $\sigma_{v,x}$.

Hence each of the previously considered isometries of V is induced by an appropriate inner-automorphism of C(V) . In particular, the unitary transvections appear much more natural from this viewpoint than when they were introduced in (III).

Actually, more can be said. By (V.5) each isometry τ of O(V) may be written as a product of symmetries. Thus, there are $x_1,\ldots,x_s$ nonisotropic vectors in V with

$$\tau = \sigma_{x_1}\sigma_{x_2}\cdots\sigma_{x_s}$$

where σ_{x_i} is a symmetry. Hence, for y in V ,

$$\begin{aligned}\tau(y) &= \sigma_{x_1}\sigma_{x_2}\cdots\sigma_{x_s}(y)\\ &= I_{x_1}I_{x_2}\cdots I_{x_s}(y)\\ &= I_{x_1x_2\cdots x_s}(y)\end{aligned}$$

and <u>each</u> isometry of (V,β) is induced by an inner-automorphism of C(V) !

Hence, to create isometries, one creates appropriate inner-automorphisms of C(V) . That is, select a homogeneous unit x in C(V) satisfying

$$I_x(V) \subset V .$$

Then $I_x|_V$ is in the orthogonal group. To see this, observe $I_x|_V$ is R-linear. Thus, it is sufficient to show $I_x|_V$ preserves the form. Suppose v_1 and v_2 are in V. Then

$$\begin{aligned}\beta(I_xv_1, I_xv_2) &= \beta(xv_1x^{-1}, xv_2x^{-1}) \\ &= \tfrac{1}{2}[(xv_1x^{-1})(xv_2x^{-1}) + (xv_2x^{-1})(xv_1x^{-1})] \\ &= \tfrac{1}{2}(xv_1v_2x^{-1} + xv_2v_1x^{-1}) \\ &= x(\tfrac{1}{2}(v_1v_2 + v_2v_1))x^{-1} \\ &= \beta(v_1, v_2) \quad .\end{aligned}$$

We summarize these results in the next theorem.

<u>(V.11)</u> <u>THEOREM</u>. Let x be a homogeneous unit of $C(V)$ and

$$I_x(w) = (-1)^{\partial x \partial w} x w x^{-1} \quad \text{for } w \text{ in } hC$$

be the inner-automorphism of $C(V)$ induced by x .

(a) If τ is in $O(V)$, then there is a homogeneous unit x in $C(V)$ with

$$\tau = I_x|_V \quad .$$

(b) In particular,

(i) $\sigma_x = I_x|_V$ for the symmetry σ_x (x in V and non-isotropic).

(ii) If $V = H \perp W$ where $H = Ru \oplus Rv$ is a hyperbolic plane, then:

$$\Delta = I_{u-v}\Big|_V ,$$

$$\Phi_\varepsilon = I_{\varepsilon uv+vu}\Big|_V , \text{ and}$$

$$\sigma_{u,x} = I_{1+\frac{1}{2}ux}\Big|_V \quad (\text{for } \beta(x,u) = 0).$$

To illustrate the remark prior to (V.11) concerning the creation of isometries, we examine an isometry of $O(V)$ recently discovered by D. G. James. The isometry is constructed from an inner-automorphism of $C(V)$ restricted to V .

Let r and s be in V and suppose $\beta(r,s) = 0$ and

$$1 + \beta(r,r)\beta(s,s)$$

is a unit in R . Let $w = 1 + \beta(r,r)\beta(s,s)$. In the Clifford algebra

$$\begin{aligned}(1 + rs)w^{-1}(1 - rs) &= w^{-1}(1 - rsrs)\\ &= w^{-1}(1 + r^2s^2)\\ &= w^{-1}w\\ &= 1 \quad .\end{aligned}$$

Thus $1 + rs$ is a unit of degree 0 with inverse $(1 + rs)^{-1} = w^{-1}(1 - rs) = w^{-1}(1 + sr)$.

Define an inner-automorphism $J_{r,s}$ of $C(V)$ by

$$J_{r,s}(x) = (1 + rs)x(1 + rs)^{-1} .$$

To show $J_{r,s}\big|_V$ is in $O(V)$, we show by direct computation that

$$J_{r,s}\big|_V : V \to V .$$

For x in V , observe that in $C(V)$

$$\begin{aligned} xsr &= 2\beta(x,s)r - sxr \\ &= 2\beta(x,s)r - [2\beta(x,r)s - srx] \\ &= 2\beta(x,s)r - 2\beta(x,r)s + srx \end{aligned}$$

and thus

$$\begin{aligned} rsxsr &= rs[2\beta(x,s)r - 2\beta(x,r)s + srx] \\ &= -2\beta(x,s)\beta(r,r)s - 2\beta(x,r)\beta(s,s)r + \beta(r,r)\beta(s,s)x \end{aligned}$$

since $rs = -sr$.

Then, for x in V ,

$$\begin{aligned} J_{r,s}(x) &= (1 + rs)x(1 + rs)^{-1} \\ &= w^{-1}(1 + rs)x(1 + sr) \\ &= w^{-1}[x + xsr + rsx + rsxsr] . \end{aligned}$$

Substituting the above formulas for xsr and $rsxsr$, a short computation yields

$$J_{r,s}(x) = x + 2w^{-1}[\beta(x,s - \beta(s,s)r)r - \beta(x,r + \beta(r,r)s)s] .$$

Hence, $J_{r,s}(V) \subset V$ and $J_{r,s}|_V$ is in $O(V)$.

The next lemma is straightforward.

<u>(V.12)</u> <u>LEMMA</u>. Let r and s be in V with $\beta(r,s) = 0$ and $w = 1 + \beta(r,r)\beta(s,s)$ a unit in R . Then the inner-automorphism

$$J_{r,s}(x) = (1 + rs)x(1 + rs)^{-1}$$

of the Clifford algebra $C(V)$ induces by restriction an isometry on V . Further,

(a) $J_{r,s}(x) = x + 2w^{-1}[\beta(x,s - \beta(s,s)r)r - \beta(x,r + \beta(r,r)s)s]$ for x in V .

(b) For Θ in $O(V)$,

$$\Theta J_{r,s}\Theta^{-1} = J_{\Theta r,\Theta s} .$$

(c) If r is isotropic, then

$$J_{r/2,s} = J_{r,s/2}$$

is the unitary transvection $\sigma_{r,s}$.

By (c) above the isometry $J_{r,s}$ gives the unitary transvections. A second situation of interest is when r is nonisotropic, i.e., $\beta(r,r)$ is a unit.

Suppose x and $\bar{x}$ are nonisotropic vectors in V of the same norm, i.e.,

$$\beta(x,x) = \beta(\bar{x},\bar{x}) .$$

We seek to construct an s with

$$J_{x,s}(\bar{x}) = x .$$

Since x is nonisotropic, $V = Rx \perp (Rx)^{\perp}$. Thus

$$\bar{x} = \alpha x + y$$

where α is in R and y is in $(Rx)^{\perp}$. For the moment s will be undetermined. Compute

$$\begin{aligned} J_{x,s}(\bar{x}) &= (\alpha x + y) + 2w^{-1}[\beta(\alpha x + y,s - \beta(s,s)x)x \\ &\qquad - \beta(\alpha x + y,x + \beta(x,x)s)s] \\ &= [\alpha + 2w^{-1}(\beta(y,s) - \alpha\beta(x,x)\beta(s,s))]x \\ &\qquad + [y - 2w^{-1}(\alpha\beta(x,x) + \beta(x,x)\beta(y,s))s] \end{aligned} .$$

The aim is to select s with $\beta(x,s) = 0$, $1 + \beta(x,x)\beta(s,s)$ a unit and such that the second term in the above sum is 0 . A candidate for s would be δy where δ is in R and is to be determined.

Certainly, if $s = \delta y$ then $\beta(x,s) = 0$. Thus, let $s = \delta y$ for some δ in R which will be determined. The δ must be selected so that the coefficient of y above will equal 0 . That is,

$$0 = 1 - 2w^{-1}\delta(\alpha\beta(x,x) + \beta(x,x)\beta(y,y)\delta)$$

$$0 = w - 2\alpha\beta(x,x)\delta - 2\beta(x,x)\beta(y,y)\delta^2 .$$

Since $\bar{x} = \alpha x + y$ and $\beta(x,x) = \beta(\bar{x},\bar{x})$,

$$\beta(x,x) = \alpha^2\beta(x,x) + \beta(y,y) .$$

Hence, using

$$w = 1 + \beta(x,x)\beta(s,s) = 1 + \delta^2\beta(x,x)\beta(y,y)$$

and

$$\beta(y,y) = \beta(x,x)(1 - \alpha^2) ,$$

the equation δ must satisfy becomes

$$0 = 1 - 2\delta\alpha\beta(x,x) - (1 - \alpha^2)\delta^2\beta(x,x)^2 .$$

Let $X = \delta\beta(x,x)$. Then

$$0 = 1 - 2\alpha X - (1 - \alpha^2)X^2 .$$

A direct computation with the quadratic formula, shows

$$X = \frac{1}{\alpha+1} \qquad \text{or} \qquad X = \frac{1}{\alpha-1} .$$

Of course, $\alpha + 1$ or $\alpha - 1$ may not be units. However, 2 is a unit and $(\alpha + 1) - (\alpha - 1) = 2$. Hence, $\alpha + 1$ or $\alpha - 1$ is a unit. Let ε denote the choice of $\alpha + 1$ or $\alpha - 1$ which gives a unit. Then, take

$$\delta = \frac{1}{\varepsilon\beta(x,x)} .$$

This choice of δ will produce a zero coefficient for y. Further,

$$\begin{aligned} w &= 1 + \beta(x,x)\beta(s,s) \\ &= 1 + \beta(x,x)\beta(y,y)\delta^2 \\ &= 1 + \beta(x,x)\delta^2(\beta(x,x)(1 - \alpha^2)) \\ &= 1 + \frac{1}{\varepsilon^2}(1 - \alpha^2) \\ &= \pm \frac{2}{\varepsilon} \end{aligned}$$

is a unit. Hence, for the above choice of s

$$J_{x,s}(\bar{x}) = \mu x .$$

It is easy to see $\mu = \pm 1$.

<u>(V.13)</u> <u>LEMMA</u>. Let x and $\bar{x}$ be nonisotropic vectors in V of the same norm, i.e., $\beta(x,x) = \beta(\bar{x},\bar{x})$. Then, for suitable s ,

$$J_{x,s}(\bar{x}) = x$$

or

$$(-J_{x,s})(\bar{x}) = x .$$

<u>(V.14)</u> <u>THEOREM</u>. (James) Let (V,β) have an orthogonal basis $\{x_1,\ldots,x_n\}$. If ϕ is in $O(V)$, then

$$\phi = J_{x_1,t_1} J_{x_2,t_2} \cdots J_{x_n,t_n} \Theta$$

for suitable $t_1,\ldots,t_n$ where

(a) $\beta(x_i,t_j) = 0$ for $i \leq j$.

(b) Θ is an isometry satisfying $\Theta(x_i) = \pm x_i$ for each i .

Proof. The proof will be by induction on $n = \dim(V)$. Let ϕ be in $O(V)$. Let $V = Rx_1 \perp (Rx_1)^{\perp}$. Suppose $\phi(x_1) = \bar{x}_1$. Then $\beta(\bar{x}_1,\bar{x}_1) = \beta(x_1,x_1)$. By (V.13) there is a t_1 with

$$x_1 = \Theta J_{x_1,t_1}(\bar{x}_1) = (\Theta J_{x_1,t_1}\phi)(x_1)$$

where $\Theta(x_1) = \pm x_1$ and $\Theta(x_i) = x_i$ for $i \geq 2$. Then, $\Theta J_{x_1,t_1}\phi : (Rx_1)^{\perp} \to (Rx_1)^{\perp}$. The result now follows by induction in a fashion similar to, say, (III.24).

We turn from the examination of a single homogeneous unit x in $C(V)$ and the inner-automorphism I_x to groups of such invertible elements.

The Clifford group $CL(V)$ of (V,β) is defined to be

$$CL(V) = \{x \text{ in } C(V) \mid x \text{ is a homogeneous unit and } I_x(V) \subseteq V\} .$$

If x is in the Clifford group $CL(V)$, then the map

$$\hat{I} : x \to I_x\big|_V$$

which carries x to the inner-automorphism I_x restricted to V determines a group morphism

$$\hat{I} : CL(V) \to O(V) .$$

Since each element τ of $O(V)$ is induced by a suitable inner-automorphism I_x for x in $CL(V)$, we have:

$$\hat{I} : CL(V) \to O(V)$$

surjective.

A second group arises from the intersection of $CL(V)$ with the homogeneous elements of degree 0 in $C(V)$, i.e., C_0. Set

$$CL_0(V) = C_0 \cap CL(V)$$

and call $CL_0(V)$ the <u>special Clifford group</u> of (V,β).

We determine the kernel of $\hat{I}$. If x is in $CL(V)$ (thus x is homogeneous) and x is in the kernel of $\hat{I}$ then

$$y = I_x(y) = (-1)^{\partial x \partial y} xyx^{-1} = (-1)^{\partial x} xyx^{-1}$$

for all y in V (∂y is thus 1). Hence, if $\partial x = 0$ then $xy = yx$ and if $\partial x = 1$ then $xy = -yx$ for all y in V. Suppose $\partial x = 0$. Then x is in C_0 and $xy = yx$ for all y. Selecting y from an orthogonal basis $\{x_1,\ldots,x_n\}$ of V, x must commute with each element of

$$\{x_1^{\varepsilon_1}\cdots x_n^{\varepsilon_n} \mid \varepsilon_i = 0,1 \text{ and } \textstyle\sum\varepsilon_i \equiv 0 \mod 2\}$$

which is a basis for C_0. Hence if $\partial x = 0$ and $I_x|_V$ = identity then x is in the center of C_0. If $\partial x = 1$ then $xx_i = -x_i x$ for all i,

$1 \le i \le n$. It is straightforward to show this case is not possible; for example, suppose $x = x_1x_2x_3$, then $xx_1 = (x_1x_2x_3)x_1 = (-1)^2 x_1x_1x_2x_3 = x_1(x_1x_2x_3) = x_1x \neq -x_1x$. More generally,

$$x = \sum \alpha_\lambda X_\lambda$$

where $X_\lambda = x_1^{\varepsilon_1} \cdots x_n^{\varepsilon_n}$ and $\sum \varepsilon_i \equiv 1 \mod 2$. Careful choice of x_i shows $xx_i = -x_i x$ will not occur. We conclude $\text{Ker}(\hat{I})$ = center of C_0 .

To determine the elements x in the center of C_0 , x must commute with $\{x_1^{\varepsilon_1} \cdots x_n^{\varepsilon_n} \mid \varepsilon_i = 0,1 \text{ and } \sum \varepsilon_i \equiv 0 \mod 2\}$. A direct computation will give part (b) of the next lemma. A similar computation will give part (a).

<u>(V.15)</u> <u>LEMMA</u>.

(a) If $\dim(V)$ is even, then the center of $C(V)$ is R . If $\dim(V)$ is odd, then the center of $C(V)$ is $R + Re$ where $e = x_1x_2 \cdots x_n$ for an orthogonal basis $\{x_1, \ldots, x_n\}$ of V .

(b) The center of C_0 is R .

Therefore, if R^* denotes the units of R , we have an exact sequence

$$1 \longrightarrow R^* \longrightarrow CL(V) \xrightarrow{\hat{I}} O(V) \longrightarrow 1 .$$

An element in $O(V)$ may be written as a product of symmetries (indeed, this is why $\hat{I}$ is surjective). Suppose τ is in $O(V)$ and

$\tau = \sigma_{y_1}\sigma_{y_2}\cdots\sigma_{y_s}$ where σ_{y_i} is a symmetry for $1 \le i \le s$. Then

$$\begin{aligned}\tau &= \sigma_{y_1}\sigma_{y_2}\cdots\sigma_{y_s} \\ &= (I_{y_1}I_{y_2}\cdots I_{y_s})|_V \\ &= I_{y_1y_2\cdots y_s}\Big|_V\end{aligned}$$

and $\hat{I} : y_1\cdots y_s \to \tau$. Suppose τ is <u>also</u> the image under $\hat{I}$ of an element w in the special Clifford group $CL_0(V)$. Thus, $I_w\Big|_V = I_{y_1y_2\cdots y_s}\Big|_V$ and

$$I_{w^{-1}y_1y_2\cdots y_s}\Big|_V = \text{identity of } O(V) .$$

Then, $w^{-1}y_1y_2\cdots y_s$ is in the kernel of $\hat{I}$ and the

$$w^{-1}y_1y_2\cdots y_s = \alpha$$

for α a unit in R . Then

$$s = \partial(w^{-1}y_1y_2\cdots y_s) \equiv 0 \mod 2 ,$$

and s is even. Replace y_1 by $\alpha^{-1}y_1$ (which is still nonisotropic) and

$$w = y_1y_2\cdots y_s$$

where s is even. We conclude that each element in the special Clifford group $CL_0(V)$ of V is a product of an even number of nonisotropic vectors from V .

Let w be in $CL_0(V)$ and suppose $w = y_1 y_2 \cdots y_s$ where y_i are non-isotropic for $1 \leq i \leq s$ and s is even. Then

$$I_w\Big|_V = \sigma_{y_1}\sigma_{y_2}\cdots\sigma_{y_s}$$

and, consequently,

$$\det(I_w\Big|_V) = \prod_i \det(\sigma_{y_i}) = (-1)^s = 1$$

since s is even. Thus,

$$\hat{I} : CL_0(V) \to SO(V)$$

and we have the exact sequence of groups

$$1 \longrightarrow R^* \longrightarrow CL_0(V) \xrightarrow{\hat{I}} SO(V) \longrightarrow 1 .$$

Let $x \to x^*$ denote the R-algebra anti-automorphism of $C(V)$ which is the identity on V and consists of reversing the order of multiplication. Utilizing the mapping properties of the Clifford algebra it can be shown $x \to x^*$ is the unique R-algebra anti-automorphism of $C(V)$ which is the identity on V, e.g., see O'Meara ([22], p.136) or Bass ([40], p.172).

Following Bass ([40], p.173), let $\mu : C(V) \to C(V)$ be the inner-automorphism of $C(V)$ satisfying

$$\mu|_V = -(\text{identity on } V) ,$$

and let $x \to \bar{x}$ denote the composition of $(\ )^*$ and μ. Thus

$x \to \bar{x}$ is an anti-automorphism $C(V) \to C(V)$ satisfying

$$\bar{x} = -x$$

for x in V. Clearly $\bar{\bar{x}} = x$ for all x in $C(V)$, since this is true for x in V.

Let $M = \{x \text{ in } C(V) \mid x\bar{x} \text{ is in } R\}$. If x and y are in M then

$$\begin{aligned}(xy)(\overline{xy}) &= xy\bar{y}\bar{x} \\ &= x\bar{x}y\bar{y}\end{aligned}$$

since $y\bar{y}$ is in R. Hence, M is a semi-group and

$$N : M \to R$$

by

$$N(x) = x\bar{x}$$

is a morphism called the norm. Note if x is in V then $x\bar{x} = -x^2 = -\beta(x,x)$.

(V.16) LEMMA. If x is in $CL(V)$ then $\bar{x}$ is in $CL(V)$ and $N : CL(V) \to R^*$ is a group morphism. If r is in R^* then $N(r) = r^2$.

Proof. Suppose x is in $CL(V)$ and y is in V. Then $I_x(y)$ is in V. By the above remark,

$$
\begin{aligned}
I_x(y) &= -(\overline{I_x(y)}) \\
&= (-1)(-1)^{\partial x}\overline{xyx^{-1}} \\
&= (-1)^{1+\partial x+1}\bar{x}^{-1}y\bar{x} \\
&= I_{\bar{x}^{-1}}(y)
\end{aligned}
$$

since $\partial x = \partial(\bar{x}^{-1})$. Thus $I_{\bar{x}}\big|_V = I_x^{-1}\big|_V$ and $\bar{x}$ is in $CL(V)$. Clearly, $x\bar{x}$ is in $Ker(\hat{I})$, i.e., $x\bar{x}$ is in R^* . The remainder is straightforward.

Define the following groups:

$$
\begin{aligned}
Pin(V) &= Ker(N : CL(V) \to R^*) \ , \\
Spin(V) &= Ker(N : CL_0(V) \to R^*) \ .
\end{aligned}
$$

Observe

$$
Spin(V) = Pin(V) \cap CL_0(V)
$$

and

$$
\begin{aligned}
Pin(V) \cap R^* &= Spin(V) \cap R^* \\
&= \{r \text{ in } R \mid r^2 = 1\} \\
&= \{\pm 1\} \ .
\end{aligned}
$$

In $CL(V)$ we have the group diagram:

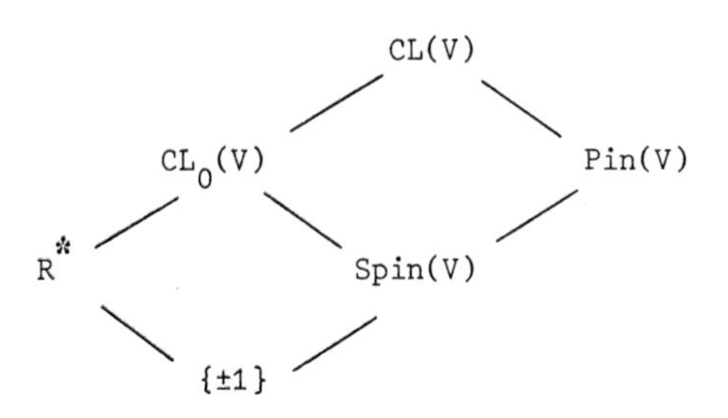

The above diagram transforms under $\hat{I}$ to:

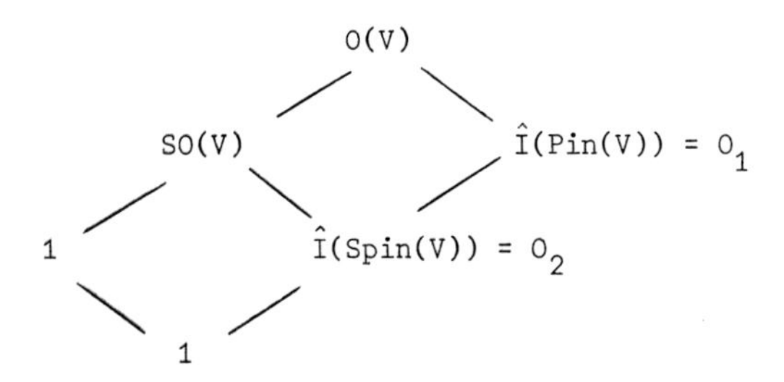

where

$$O_1 = \hat{I}(Pin(V))$$
$$O_2 = \hat{I}(Spin(V)) .$$

A straightforward calculation shows

$$\{\pm 1\} = Ker(\hat{I} : Pin \to O_1)$$
$$= Ker(\hat{I} : Spin \to O_2) .$$

The group O_2 is often called the reduced orthogonal group of V.[1] Our discussion is summarized in the following diagram of Bass [40]:

---

[1] See (V.28) and (V.29).

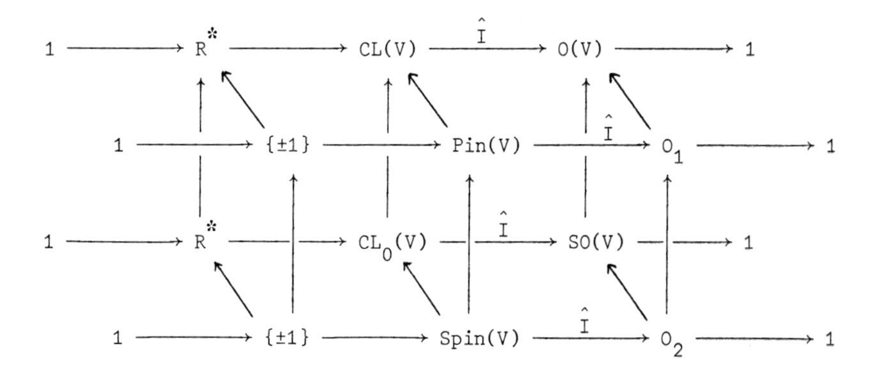

(E) THE CONGRUENCE-COMMUTATOR GROUPS

Let (V,β) be a symmetric inner product space over the local ring R. We do not initially place any assumption on the hyperbolic rank of V. We do assume $\dim(V) \geq 3$.

(V.17) PROPOSITION. Suppose the dimension of V is ≥ 3. Then the center of $O(V)$ is $\{\pm I\}$.

Proof. Let $V = Rx \perp Ry \perp Rz \perp W$ where x, y and z are members of an orthogonal basis for V (note $\dim(V) \geq 3$). Clearly $\{\pm I\} \subset \mathrm{Center}(O(V))$. It is necessary to show that if τ is in $\mathrm{Center}(O(V))$ then $\tau = \pm I$. If s is nonisotropic in V then, since $\sigma_s\tau = \tau\sigma_s$

for the symmetry σ_s , we have $\tau : (Rs)^{\perp} \to (Rs)^{\perp}$ and thus $\tau : Rs \to Rs$. Hence $\tau(s) = \alpha_s s$ for some unit α_s in R . In particular,

$$\tau x = \alpha_x x$$
$$\tau y = \alpha_y y$$
$$\tau z = \alpha_z z \quad .$$

Clearly $\alpha_x^2 = \alpha_y^2 = \alpha_z^2 = 1$. It is necessary to show $\alpha_x = \alpha_y = \alpha_z$.

Suppose that under the canonical ring morphism $\Pi_m : R \to R/m$,

$$\Pi_m\beta(x + y + z, x + y + z) \neq 0 \ .$$

Then $x + y + z$ is nonisotropic and, as above,

$$\begin{aligned} \alpha_x x + \alpha_y y + \alpha_z z &= \tau x + \tau y + \tau z \\ &= \tau(x + y + z) \\ &= \alpha_{x+y+z}(x + y + z) \end{aligned}$$

for some α_{x+y+z} in R . Since x , y and z are R-free we have $\alpha_x = \alpha_y = \alpha_z = \alpha_{x+y+z}$.

On the other hand, if

$$\begin{aligned} 0 &= \Pi_m\beta(x + y + z, x + y + z) \\ &= \Pi_m\beta(x,x) + \Pi_m\beta(y,y) + \Pi_m\beta(z,z) \end{aligned}$$

then, since x , y and z are nonisotropic,

$$0 \neq \Pi_m\beta(x,x) + \Pi_m\beta(y,y)$$
$$= \Pi_m\beta(x + y, x + y)$$

and

$$0 \neq \Pi_m\beta(x,x) + \Pi_m\beta(z,z)$$
$$= \Pi_m\beta(x + z, x + z) \quad .$$

An argument similar to the above gives $\alpha_x = \alpha_y$ and $\alpha_x = \alpha_z$, finishing the proof.

Suppose A is an ideal of R with $A \neq R$. The ring morphism $\Pi_A : R \to R/A$ induces a surjective R-morphism $\Pi_A : (V,\beta) \to (V/AV,\beta_A)$ of symmetric inner product spaces where $\beta_A(\Pi_A x,\Pi_A y) = \Pi_A\beta(x,y)$. In turn, this gives a surjective (see (V.6)) group morphism $\lambda_A : O(V) \to O(V/AV)$. Recall

$$SO(V) = \{\sigma \text{ in } O(V) \mid \det(\sigma) = 1\} \quad .$$

Define

$$O(V,A) = \{\sigma \text{ in } O(V) \mid \lambda_A\sigma \text{ is in } \mathrm{Center}(O(V/AV))\} \ .$$

Since $\dim(V) \geq 3$, (V.17) implies

$$O(V,A) = \{\sigma \text{ in } O(V) \mid \lambda_A\sigma = \pm I\} \ .$$

Let

$$SO(V,A) = SO(V) \cap O(V,A) \ .$$

The group $O(V,A)$ is the _congruence subgroup of_ $O(V)$ _of level_ A and the group $SO(V,A)$ is the _special congruence subgroup of_ $O(V)$ _of level_ A .

If $A = R$, hence $V/RV = 0$, let $O(V/RV) = I$ and $SO(V/RV) = I$. Further, let

$$O(V,R) = O(V)$$

$$O(V,0) = \text{Center } O(V) ,$$

$$SO(V,R) = SO(V) , \text{ and}$$

$$SO(V,0) = SO(V) \cap \text{Center } O(V) .$$

Let

$$\Omega(V) = [O(V),O(V)]$$

denote the _commutator subgroup of_ $O(V)$, i.e., the subgroup generated by all $g^{-1}h^{-1}gh$ for g and h in $O(V)$. Define the _mixed commutator subgroup of level_ A to be

$$\Omega(V,A) = [O(V),O(V,A)]$$

for an ideal A , i.e., the subgroup generated by all $g^{-1}h^{-1}gh$ for g in $O(V)$ and h in $O(V,A)$. Observe

$$\Omega(V,A) \leq O(V,A) .$$

Indeed,

$$\Omega(V,A) \leq SO(V,A) .$$

Suppose V has hyperbolic rank ≥ 1. Thus assume $V = H \perp W$ where $H = Ru \oplus Rv$, $\beta(u,v) = 1$ and $\beta(u,u) = \beta(v,v) = 0$. Let

$$EO(V)$$

denote the subgroup of $O(V)$ generated by all the unitary transvections $\sigma_{u,x}$ and $\sigma_{v,y}$ where x and y are in W. For an ideal A of R, let

$$EO(V,A)$$

denote the $EO(V)$-normal subgroup of $O(V,A)$ generated by all unitary transvections $\sigma_{u,x}$ and $\sigma_{v,y}$ where x and y are in W and $\Pi_A x = \Pi_A y = 0$, i.e., the order $O(x)$ and the order $O(y)$ of x and y are in A. We call $EO(V)$ the _Eichler subgroup_ of $O(V)$ and $EO(V,A)$ the _Eichler subgroup of level_ A. Observe

$$EO(V,A) \leq O(V,A)$$

and $EO(V,A)$ is the subgroup generated by all $\Theta\sigma_{u,x}\Theta^{-1}$ and $\Theta\sigma_{v,y}\Theta^{-1}$ for Θ in $EO(V)$ and $O(x) \subset A$, $O(y) \subset A$.

Our first purpose will be to obtain some characterizations of $\Omega(V,A)$. Preliminary to this we make some additional observations on symmetries.

Suppose P is a plane in V. Recall P is _non-singular_ (_nonisotropic_ or _regular_) if the restriction $\beta|_{P\times P}$ is an inner product.

Suppose τ is in $\Omega(V,A) = [O(V),O(V,A)]$ and A is not equal to R. By examining the map $\lambda_A : O(V) \to O(V/AV)$ on the generators of $\Omega(V,A)$, i.e., on the commutators $\mu\rho\mu^{-1}\rho^{-1}$ where μ is in $O(V)$ and ρ is in $O(V,A)$, it is easy to see that $\lambda_A(\tau) = I$. By Corollary (V.7), τ is a product of elements of the form $\sigma_1\sigma_2$ where σ_1 and σ_2 are symmetries and $\lambda_A(\sigma_1\sigma_2) = I$. Thus, we examine a product $\sigma_1\sigma_2$ of symmetries satisfying $\lambda_A(\sigma_1\sigma_2) = I$.

(V.18) LEMMA. Let A be an ideal of R and $\dim(V) \geq 3$. Suppose σ and $\bar{\sigma}$ are symmetries satisfying

$$\lambda_A(\sigma\bar{\sigma}) = I$$

Then there are symmetries $\sigma_1,\ldots,\sigma_s$ ($s \leq n$) and non-singular planes $P_1,\ldots,P_{s-1}$ with

(a) $\sigma = \sigma_1$, $\bar{\sigma} = \sigma_s$;

(b) $\sigma\bar{\sigma} = \prod_{i=1}^{s-1} (\sigma_i\sigma_{i+1})$;

(c) $\sigma_i\sigma_{i+1}$ = identity on $P_i^{\perp}$ for $1 \leq i \leq s-1$;

(d) $\lambda_A(\sigma_i\sigma_{i+1}) = I_A$.

Proof. Suppose initially that $A \neq R$. Let $\sigma = \sigma_x$ and $\bar{\sigma} = \sigma_{\bar{x}}$ for nonisotropic vectors x and $\bar{x}$. Extend x to an orthogonal basis $\{x_1,x_2,\ldots,x_n\}$ of V where $x = x_1$. Define

$$y_1 = x_1 ,$$
$$y_2 = \alpha_1x_1 + \alpha_2x_2 ,$$

$$\vdots$$

$$y_n = \alpha_1 x_1 + \alpha_2 x_2 + \cdots + \alpha_n x_n$$

where $\bar{x} = \alpha_1 x_1 + \alpha_2 x_2 + \cdots + \alpha_n x_n$. Hence $y_n = \bar{x}$.

Since $\lambda_A(\sigma\bar{\sigma}) = \lambda_A(\sigma_x \sigma_{\bar{x}}) = \sigma_{\Pi_A x}\sigma_{\Pi_A \bar{x}} = I$, we have $\sigma_{\Pi_A \bar{x}}^{-1} = \sigma_{\Pi_A x}$. Thus $\Pi_A x = (\text{unit})\, \Pi_A \bar{x}$ and, hence, α_1 is a unit and $\alpha_2, \ldots, \alpha_n$ are in $A \subset m$. Thus, each y_i is unimodular and since

$$\beta(y_i, y_i) = \sum_{j=1}^{i} \alpha_j^2 \beta(x_j, x_j) = \text{unit} + (\text{element of } m) ,$$

we have y_i nonisotropic for $1 \le i \le n$.

Set $\sigma_i = \sigma_{y_i}$ for $1 \le i \le n$. Then $\sigma_1 = \sigma$ and $\sigma_n = \bar{\sigma}$. Further,

$$\begin{aligned} \sigma\bar{\sigma} &= (\sigma_1\sigma_2)(\sigma_2\sigma_3)\cdots(\sigma_{n-1}\sigma_n) \\ &= \Pi\sigma_i\sigma_{i+1} \end{aligned}$$

and

$$\begin{aligned} \lambda_A \sigma_i \sigma_{i+1} &= \lambda_A \sigma_{y_i} \sigma_{y_{i+1}} \\ &= \sigma_{\Pi_A y_i} \sigma_{\Pi_A y_{i+1}} \\ &= \sigma_{\Pi_A y_i} \sigma_{\Pi_A y_i} \qquad (\ \Pi_A y_i = \Pi_A y_{i+1}\ ) \\ &= I \end{aligned} .$$

The planes P_i are given by

$$P_i = Ry_i \oplus Rx_{i+1} .$$

Then, it is easy to check that P_i is a non-singular plane and

$$\sigma_i\sigma_{i+1}\big|_{P_i^\perp} = \text{identity} .$$

Suppose $A = R$, $\sigma = \sigma_x$ and $\bar{\sigma} = \sigma_{\bar{x}}$. Since $\dim(V) \geq 3$, it is easy to select a nonisotropic y in V such that $P_1 = Rx \oplus Ry$ and $P_2 = R\bar{x} \oplus Ry$ are non-singular planes. For example, reduce the space V modulo mV , make the selection in the R/m-space V/mV and then choose an appropriate pre-image. Setting $\sigma_x = \sigma_1$, $\sigma_y = \sigma_2$ and $\sigma_{\bar{x}} = \sigma_3$,

$$\sigma\bar{\sigma} = (\sigma_1\sigma_2)(\sigma_2\sigma_3)$$

completing the proof.

The above products of symmetries $\sigma_i\sigma_{i+1}$ have the property that $\sigma_i\sigma_{i+1}\big|_{P_i^\perp} = \text{identity}$, i.e., their action is restricted to a plane (non-singular). Further $\det(\sigma_i\sigma_{i+1}) = 1$. Thus, $\sigma_i\sigma_{i+1}$ is called a <u>(non-singular) plane rotation</u>. Similarly, Δ and Φ_ε are (non-singular) plane rotations, while the unitary transvections $\sigma_{u,x}$ and $\sigma_{v,x}$ are <u>(singular) plane rotations</u>. The lemma shows that each element of $\Omega(V,A)$ may be realized as a sequence of non-singular plane rotations.

(V.19) THEOREM. (Characterization of $\Omega(V,A)$) Let $\dim(V) \geq 3$ and A be an ideal of R.

(a) (Klingenberg) Then

$$\Omega(V,A) = [SO(V), SO(V,A)] .$$

Further, $\Omega(V,A)$ is the normal subgroup generated by

$$\{(\sigma\bar{\sigma})^2 \mid \sigma \text{ and } \bar{\sigma} \text{ are symmetries and } \lambda_A(\sigma\bar{\sigma}) = I\} .$$

(b) (James) If the hyperbolic rank of V is ≥ 1 and 3 is a unit[1] in R then

$$\Omega(V,A) = EO(V,A) .$$

Proof. We begin with a proof of (a). Suppose that τ is in $O(V,A)$ and x is a nonisotropic vector in V. Then, as noted previously, $\tau x + \varepsilon x$ is nonisotropic for either $\varepsilon = 1$ or $\varepsilon = -1$. Then, letting $\bar{\sigma}$ denote the symmetry $\sigma_{\tau x+\varepsilon x}$,

$$\bar{\sigma}\tau x = -\varepsilon x .$$

---

[1] The hypothesis that 3 is a unit may be replaced by the less restrictive condition $|R/m| \geq 4$, i.e., R/m is not the field of 3 elements. One has to only be slightly more careful in the choice of units in the proof of (b). See James [87].

Hence,

$$\tau x = -\varepsilon\bar{\sigma}x$$

and the symmetry

$$\sigma_{\tau x} = \sigma_{-\varepsilon\bar{\sigma}x} = \sigma_{\bar{\sigma}x} .$$

Then,

$$\begin{aligned}\sigma_x\tau\sigma_x^{-1}\tau^{-1} &= \sigma_x\tau\sigma_x\tau^{-1} = \sigma_x\sigma_{\tau x}\\ &= \sigma_x\sigma_{\bar{\sigma}x} = \sigma_x(\bar{\sigma}\sigma_x\bar{\sigma}^{-1})\\ &= (\sigma_x\bar{\sigma})(\sigma_x\bar{\sigma}) = (\sigma_x\bar{\sigma})^2 .\end{aligned}$$

Further,

$$\begin{aligned}\lambda_A(\sigma_x\bar{\sigma}) &= (\lambda_A\sigma_x)(\lambda_A\sigma_{\tau x+\varepsilon x})\\ &= \sigma_{\Pi_A x}\sigma_{2\varepsilon\Pi_A x} = I .\end{aligned}$$

Let $\rho\tau\rho^{-1}\tau^{-1}$ be a generator of $\Omega(V,A) = [O(V),O(V,A)]$ where ρ is in $O(V)$ and τ is in $O(V,A)$. Since $O(V)$ is generated by symmetries, let

$$\rho = \sigma_x\bar{\rho} \qquad \text{where } \sigma_x \text{ is a symmetry} .$$

Then,

$$\begin{aligned}\rho\tau\rho^{-1}\tau^{-1} &= \sigma_x\bar{\rho}\tau\bar{\rho}^{-1}\sigma_x^{-1}\tau^{-1}\\ &= \sigma_x(\bar{\rho}\tau\bar{\rho}^{-1}\tau^{-1})(\tau\sigma_x^{-1}\tau^{-1}\sigma_x)\sigma_x .\end{aligned}$$

Since $\bar{\rho}$ has one fewer symmetries than ρ, an induction argument shows $\bar{\rho}\tau\bar{\rho}^{-1}\tau^{-1}$ has the desired form. Certainly, $\tau\sigma_x^{-1}\tau^{-1}\sigma_x$ has the desired form.

Therefore, if G_A denotes the normal subgroup generated by all $(\sigma\bar{\sigma})^2$ with σ and $\bar{\sigma}$ being symmetries which satisfy $\lambda_A(\sigma\bar{\sigma}) = I$, then

$$G_A \geq \Omega(V,A) = [O(V),O(V,A)] \geq [SO(V),SO(V,A)] .$$

It remains to show

$$G_A \leq [SO(V),SO(V,A)] .$$

Let σ and $\bar{\sigma}$ be symmetries with $\lambda_A(\sigma\bar{\sigma}) = I$. By the lemma,

$$\sigma\bar{\sigma} = \prod_{i=1}^{s-1} \sigma_i\sigma_{i+1}$$

where σ_i and σ_{i+1} are symmetries and $\sigma_i\sigma_{i+1}$ is a "plane rotation" on a non-singular plane P_i. Set $\tau = \prod_{i=2}^{s-1} \sigma_i\sigma_{i+1}$. Then

$$\begin{aligned}(\sigma\bar{\sigma})^2 &= (\prod_{i=1}^{s-1} \sigma_i\sigma_{i+1})^2 \\ &= (\sigma_1\sigma_2\tau)^2 \\ &= (\sigma_1\sigma_2)^2[(\sigma_2\sigma_1)\tau(\sigma_2\sigma_1)^{-1}\tau^{-1}]\tau^2 .\end{aligned}$$

The expression in " [ ] " is in $[SO(V),SO(V,A)]$ while the term τ^2 has one fewer plane rotations than $(\sigma\bar{\sigma})^2$. An induction argument would finish the proof if $(\sigma_1\sigma_2)^2$ were in $[SO(V),SO(V,A)]$. Thus, suppose P is a non-singular plane containing nonisotropic vectors x and y and consider $(\sigma_x\sigma_y)^2$. By (III.6), $P^\perp$ is non-singular and $V = P \perp P^\perp$. Hence $\dim(P^\perp) = n - 2 \geq 1$ since $\dim(V) \geq 3$. Select a nonisotropic vector z in $P^{\perp}$ and consider $W = P \perp Rz$. Observe

$$\sigma_x = \sigma_x\big|_W \perp I\big|_{W^{\perp}} ,$$

$$\sigma_y = \sigma_y\big|_W \perp I\big|_{W^{\perp}} .$$

Then $\sigma_x\sigma_y = \tau\bar{\tau}$ where

$$\tau = (-\sigma_x)\big|_W \perp I\big|_{W^{\perp}} ,$$

$$\bar{\tau} = (-\sigma_y)\big|_W \perp I\big|_{W^{\perp}} .$$

But τ and $\bar{\tau}$ are in $SO(V)$ since $\det(\tau) = \det(\bar{\tau}) = 1$. Then

$$(\sigma_x\sigma_y)^2 = (\tau\bar{\tau})^2 = \tau(\bar{\tau}\tau)\tau^{-1}(\bar{\tau}\tau)^{-1}$$

and $(\sigma_x\sigma_y)^2$ is in $[SO(V),SO(V,A)]$.

To prove part (b) we need the following technical lemma. Over fields this lemma was established by Eichler [16] and was extended to local rings by James [87]. It may be verified directly by checking the images of u , v , and x in W under the action of the right and left sides of the equation.

(V.20) LEMMA. (Eichler-James) Suppose V has hyperbolic rank ≥ 1 and $V = (Ru \oplus Rv) \perp W$ for a hyperbolic plane $H = Ru \oplus Rv$. If α and δ are in R and x is in W with

$$\eta = 1 - \alpha\delta \frac{\beta(x,x)}{2}$$

a unit, then

$$\sigma_{v,\alpha x}\sigma_{u,\delta x} = \sigma_{u,\eta^{-1}\delta x}\sigma_{v,\alpha\eta x}{}^{\Phi}{}_{\eta^{-2}} .$$

We begin the proof of (b) of (V.19). We want to show the Eichler group $EO(V,A)$ of level A is equal to $\Omega(V,A) = [O(V),O(V,A)]$ if 3 is a unit and hyperbolic rank of $V \geq 1$. We assume the notation of the lemma.

Let x be in W with $O(x) \subset A$. Then, by (III.16),

$$\begin{aligned} \Phi_3\sigma_{u,x/2}\Phi_3^{-1}\sigma_{u,x/2}^{-1} &= \Phi_3\sigma_{u,x/2}\Phi_{3^{-1}}\sigma_{u,-x/2} \\ &= \sigma_{u,x} . \end{aligned}$$

A similar expression is available for $\sigma_{v,x}$. Further, $\lambda_A\sigma_{u,x} = I$. Thus, $\sigma_{u,x}$ is in $\Omega(V,A) = [O(V),O(V,A)]$ and

$$EO(V,A) \subset \Omega(V,A) .$$

It is now necessary to show $\Omega(V,A) \subset EO(V,A)$. We utilize the results in (III), in particular, Section (D).

Suppose $A \neq R$. By (III.25), each element ϕ in $O(V,A)$ can be written as

$$\phi = \pm\Phi_\varepsilon\psi$$

where ε is a unit with $\varepsilon \equiv 1 \mod A$ and ψ is in $EO(V,A)$. (Actually, (III.25) only gives the generators. However, it is easy to check that ϕ has the above form.)

By part (a) of this theorem, a generator of $\Omega(V,A)$ has the form $(\sigma\bar{\sigma})^2$ where σ and $\bar{\sigma}$ are symmetries with $\lambda_A(\sigma\bar{\sigma}) = I$. Hence $\sigma\bar{\sigma}$ is in $O(V,A)$. By the above remark,

$$\sigma\bar{\sigma} = \pm\Phi_\varepsilon\psi$$

and

$$(\sigma\bar{\sigma})^2 = \Phi_\varepsilon\psi\Phi_\varepsilon\psi .$$

By (III.16), $\psi\Phi_\varepsilon = \Phi_\varepsilon\bar{\psi}$ for $\bar{\psi}$ in $EO(V,A)$. Hence

$$(\sigma\bar{\sigma})^2 = \Phi_\varepsilon^2\rho$$

for ρ in $EO(V,A)$.

Hence, we are done if Φ_ε^2 is in $EO(V,A)$ when $\varepsilon \equiv 1 \mod A$.

Suppose $\varepsilon \equiv 1 \mod A$ where ε is a unit. Select x in W with x nonisotropic. In Lemma (V.20) take $\alpha = (1 - \varepsilon)\frac{1}{\beta(x,x)}$ and $\delta = 2$. Then

$$\eta = 1 - (1 - \varepsilon)\frac{1}{\beta(x,x)}\beta(x,x) = \varepsilon .$$

The lemma then gives $\Phi_{\varepsilon^{-2}}$ (likewise $\Phi_{\varepsilon^2} = \Phi_\varepsilon^2$) as a product of elements from $EO(V,A)$.

A proof similar to the above will handle the special case of $A = R$ (for example, see James [87], pp.259-260).

This completes the theorem.

<u>(V.21)</u> <u>COROLLARY</u>. Let $\dim(V) \geq 3$.

(a) $\Omega(V) = [SO(V),SO(V)]$ and $\Omega(V)$ contains the square of each element of $SO(V)$. Further, $SO(V)/\Omega(V)$ is commutative since every element has order ≤ 2 .

(b) If hyperbolic rank of $V \geq 1$ and 3 is a unit, then

$$EO(V) = \Omega(V) .$$

Let x be in V . By the <u>order</u> of x , denoted $O(x)$, we mean the smallest ideal A in R with $\Pi_A x = 0$. (See (II.A).) If σ is in $O(V)$, then the <u>order</u> of σ , denoted $O(\sigma)$, is the smallest ideal A of R satisfying

$$\lambda_A\sigma \text{ is in the center } O(V/AV) ,$$

i.e., σ is in $O(V,A)$ or $\lambda_A\sigma = \pm I$.

Let G be a subgroup of $O(V)$. The <u>order</u> of G , denoted $O(G)$, is the smallest ideal A satisfying $G \leq O(V,A)$, i.e., $\lambda_A G \leq \text{Center } O(V/AV)$. Note $O(G)$ is generated by the ideals $O(\sigma)$ as σ ranges over the elements of G .

(V.22) COROLLARY. Let $\dim(V) \geq 3$. The group morphism $\lambda_A : \Omega(V) \to \Omega(V/AV)$ is surjective. The order of $\Omega(V,A)$ is A .

(F) THE CONGRUENCE-COMMUTATOR GROUPS: HYPERBOLIC RANK ≥ 1

This section continues a study of the subgroups introduced in the previous section. We assume that the hyperbolic rank of V is ≥ 1 . Some results on the congruence-commutator groups for local rings may still be obtained without assumption on non-zero hyperbolic rank, e.g., see Klingenberg [90]; however, the sharpest theory is available if hyperbolic rank ≥ 1 . Indeed, without this assumption even the general theory of the commutator subgroups of the orthogonal group over a field is not clearly understood.

For this section we assume the following:

(a) 2 and 3 are units in R ,

(b) $\dim(V) \geq 3$,

(c) the hyperbolic rank of V is ≥ 1 .

Consequently,

$$V = H \perp W$$

where $H = Ru \oplus Rv$, $\beta(u,v) = 1$, and $\beta(u,u) = \beta(v,v) = 0$. Then, by (V.19), $\Omega(V,A) = EO(V,A)$.

Observe u (and v) is an isotropic vector in V. Conversely, by (III.14), if x is isotropic in a symmetric inner product space then that space has hyperbolic rank ≥ 1. For this reason, if a symmetric inner product space has hyperbolic rank ≥ 1, then the space is said to have isotropic vectors, to have isotropy, to have isotropic lines, or to be isotropic. Generally, we will continue with the terminology of "hyperbolic rank."

To examine $\Omega(V,A)$ from a different viewpoint, we introduce an approach of Tamagawa [143].

Let A be an ideal of R and AW be the R-submodule

$$AW = \{ \sum_{\text{finite}} a_i w_i \mid a_i \text{ in } A, \ w_i \text{ in } W\} .$$

If x is in AW then $O(x) \subset A$. On the other hand, AW is the collection of all vectors x in W with $O(x) \subset A$.

Consider the map

$$\Sigma_{u,A} : AW \to O(V)$$

defined by

$$\Sigma_{u,A}(x) = \sigma_{u,x} .$$

Using (III.16) it is easy to see $\Sigma_{u,A}$ is an injective R-module morphism of AW into $O(V)$. Indeed, $Im(\sum_{u,A}) \subset EO(V,A)$. This places a large R-submodule of W in $O(V)$.

If $\bar{u}$ is also isotropic in V then, by (III.17), there is a τ in $O(V)$ with $\tau u = \bar{u}$. Thus, for some τ in $O(V)$, since $\tau\sigma_{u,x}\tau^{-1} = \sigma_{\tau u,\tau x}$, we have

$$\tau(Im(\sum_{u,A}))\tau^{-1} = Im(\sum_{\bar{u},A}) .$$

<u>(V.23)</u> <u>LEMMA</u>. (For the above setting) The group $O(V)$ acts transitively on the set of submodules

$$\{Im(\sum_{\bar{u},A}) \mid \bar{u} \text{ isotropic}\}$$

of $O(V)$ under conjugation where A is an ideal of R.

By the lemma, the EO(V)-normal subgroup generated by $\cup\, Im(\sum_{u,A})$ where the union extends over isotropic u in V, is precisely $EO(V,A)$. Thus, if $\langle Im(\sum_{u,A})\rangle$ denotes the EO(V)-normal subgroup generated by $\cup\, Im(\sum_{u,A})$ then

$$\Omega(V,A) = \langle Im(\sum_{u,A})\rangle .$$

We employ the above observations to determine certain "higher" commutator-congruence subgroups of $O(V)$.

Define

$$C_A^1 O(V) = [O(V,A), O(V)] ,$$
$$C_A^2 O(V) = [C_A^1 O(V), O(V)] ,$$

and, for $i \geq 2$

$$C_A^{i+1} O(V) = [C_A^i O(V), O(V)] .$$

The next result was proven for fields by Tamagawa [143]. Note $C_A^1 O(V) = \Omega(V,A)$.

(V.24) THEOREM. For each $i \geq 1$,

$$C_A^i O(V) = \Omega(V,A) .$$

Proof. It suffices to show $C_A^2 O(V) = \Omega(V,A)$. We have

$$C_A^2 O(V) = [\Omega(V,A), O(V)] \leq \Omega(V,A) .$$

Thus, it is necessary to show $\Omega(V,A) \leq C_A^2 O(V)$. Since $C_A^2 O(V)$ is $EO(V)$-normal, it will suffice to show $\mathrm{Im}(\sum_{u,A}) \subseteq C_A^2 O(V)$, because

$$\Omega(V,A) = EO(V,A) = \langle \mathrm{Im}(\textstyle\sum_{u,A}) \rangle \leq C_A^2 O(V) .$$

Since 2 is a unit, Φ_2 is in $O(V)$. Let $\sigma_{u,x}$ be in $EO(V,A)$. Then

$$\begin{aligned}\Phi_2^2\sigma_{u,x}\Phi_2^{-2}\sigma_{u,x}^{-1} &= \sigma_{4u,x}\sigma_{u,-x} \\ &= \sigma_{u,4x}\sigma_{u,-x} \\ &= \sigma_{u,3x}\end{aligned}$$

is in $C_A^2O(V)$. But 3 is a unit. Thus if x is any element in AW then $\frac{1}{3}x$ is in AW and hence

$$\sigma_{u,3(\frac{1}{3}x)} = \sigma_{u,x} \text{ is in } C_A^2O(V) .$$

Therefore $\mathrm{Im}(\sum_{u,A})$ is in $C_A^2O(V)$.

<u>(V.25)</u> <u>COROLLARY.</u> Let $A = R$ and $C_R^iO(V) = C^iO(V)$.

(a) Then

$$C^iO(V) = \Omega(V)$$

for $i \geq 1$.

(b) Define

$$D^1O(V) = [O(V),O(V)]$$

and

$$D^{i+1}O(V) = [D^iO(V),D^iO(V)]$$

for $i \geq 1$. The set $\{D^iO(V) \mid i = 1,2,3,\ldots\}$ is the <u>derived</u> <u>chain</u> for $O(V)$. Then

$$D^iO(V) = \Omega(V)$$

for $i \geq 1$.

Proof. Part (a) follows directly from (V.24). To show (b), it suffices to show

$$D^2O(V) = [D^1O(V),D^1O(V)] = [\Omega(V),\Omega(V)]$$

is precisely $\Omega(V)$. Clearly $D^2O(V) \subseteq \Omega(V)$. As in the proof of (V.24) it is necessary to show $\mathrm{Im}(\sum_{u,R}) \subset D^2O(V)$. From the last remarks in Section (C), we have

$$\Phi_2 = \sigma_{u-v}\sigma_{u-2v} .$$

Then, by (V.19), Φ_2^2 is in $\Omega(V)$. The proof now follows in a fashion similar to the proof of (V.24).

We remark again that the assumption that 3 is a unit can often be replaced by the assumption that R/m is not the field of 3 elements. This is true for (V.24) if we are more careful about the choice of units.

If no assumption is made on hyperbolic rank ≥ 1 , then things become bad. O'Meara shows ([109], (5.5)) that for each integer $n \geq 2$ there is a field of characteristic not 2 and a symmetric inner product space U with hyperbolic rank $= 0$ and $\dim(U) = n$ over the field such that $\Omega(V) \neq [\Omega(V),\Omega(V)]$.
The proof of (V.25)(b) utilizes the fact that Φ_2^2 is actually in $\Omega(V)$. Thus, in the proof of (V.24), O(V) may be replaced by $\Omega(V)$ since O(V) was only used to obtain Φ_2^2 . That is, clearly

$$[\Omega(V,A),\Omega(V)] \leq [\Omega(V,A),SO(V)] \leq [\Omega(V,A),O(V)]$$

where $\Omega(V,A) = [\Omega(V,A),O(V)]$. The proof of (V.24) now shows

$$EO(V,A) = \Omega(V,A) \leq [\Omega(V,A),\Omega(V)] \ .$$

<u>(V.26)</u> <u>COROLLARY</u>.

$$\Omega(V,A) = [\Omega(V,A),\Omega(V)] = [\Omega(V,A),SO(V)] = [\Omega(V,A),O(V)] \ .$$

Since $\Omega(V) = \Omega(V,R)$, the above result may be written as

$$\Omega(V,A) = \Omega(V,RA) = [\Omega(V,A),\Omega(V,R)] \ .$$

D. G. James has asked the question, "For ideals A and B of R , does

$$\Omega(V,AB) = [\Omega(V,A),\Omega(V,B)] \ ?"$$

We compute next the centralizer of $\Omega(V)$ in $O(V)$. Suppose ϕ is in $O(V)$ and $\phi\sigma = \sigma\phi$ for all σ in $\Omega(V)$. Suppose also

$$\phi(u) = \alpha u + \delta v + y \ .$$

Let x be unimodular in W . Then $\sigma_{u,x}\phi = \phi\sigma_{u,x}$ and, consequently,

$$\begin{aligned}\alpha u + \delta v + y &= \phi(u)\\ &= \phi\sigma_{u,x}(u)\\ &= \sigma_{u,x}\phi(u)\\ &= \sigma_{u,x}(\alpha u + \delta v + y)\\ &= [\alpha + \beta(x,y) - \frac{1}{2}\delta\beta(x,x)]u + \delta v + (y - \delta x)\ .\end{aligned}$$

Thus $y = y - \delta x$ and $\delta x = 0$. But x is unimodular, hence $\delta = 0$. Comparing coefficients of u , we have $\beta(x,y) = 0$. Since $\beta|_{W\times W}$ is an inner product and x is any unimodular vector, e.g., elements of a basis of W , we have $y = 0$. Hence $\phi(u) = \alpha u$. Similarly $\phi(v) = \bar{\alpha}v$. Thus $\phi(H) \subseteq H$ and consequently $\phi(W) \subseteq W$. The proof of (III.22) now shows $\phi = \pm I$.

For X a subset of $O(V)$, let

$$C_{O(V)}(X) = \{\phi \text{ in } O(V) \mid \phi\sigma = \sigma\phi \text{ for all } \sigma \text{ in } X\}$$

denote the <u>centralizer</u> <u>of</u> X <u>by</u> $O(V)$.

<u>(V.27)</u> <u>PROPOSITION</u>. In $O(V)$,

$$C_{O(V)}(SO(V)) = C_{O(V)}(\Omega(V)) = \{\pm I\}\ .$$

Further,

$$\text{Center } SO(V) = SO(V) \cap \text{Center } O(V)\ ,$$
$$\text{Center } \Omega(V) = \Omega(V) \cap \text{Center } O(V)\ .$$

Proof. By the above discussion

$$\{\pm I\} \subseteq C_{O(V)}(SO(V)) \subseteq C_{O(V)}(\Omega(V)) = \{\pm I\} .$$

Letting ϕ be in the center of $SO(V)$ or $\Omega(V)$, respectively, in the above discussion determines the centers of $SO(V)$ and $\Omega(V)$.

Section (D) introduced a natural norm $N : CL(V) \to R^*$ taking the Clifford group $CL(V)$ into the group of units of R. We extend this in a natural fashion to the "spinor norm" on $O(V)$.

If τ is in $O(V)$ then two elements x and y of $CL(V)$ with

$$I_x\big|_V = I_y\big|_V = \tau$$

differ by a factor from R^*. Consequently, the norms, see (V.16), differ by a factor which is a square of a unit. Thus, we may define a group morphism

$$SN : O(V) \to R^*/R^{*2}$$

by

$$SN(\tau) = N(x)R^{*2}$$

where $I_x\big|_V = \tau$ for x in $CL(V)$. This morphism is called the spinor norm.

The image of SN lies in an Abelian group. Thus $\Omega(V)$ is contained in the kernel of SN .

As Artin ([1], p.194) points out, the map SN is not quite satisfactory. If the symmetric inner product β is replaced by $\alpha\beta$ for α a unit of R , then the orthogonal group remains unchanged, i.e., $\beta(x,y) = \beta(\sigma x,\sigma y)$ if and only if $(\alpha\beta)(x,y) = (\alpha\beta)(\sigma x,\sigma y)$ for σ in GL(V) . However, if $\tau = \sigma_{x_1}\sigma_{x_2}\ldots\sigma_{x_s}$ where σ_{x_i} is a symmetry then, relative to $\alpha\beta$,

$$\begin{aligned} N(\tau) &= x_1x_2\ldots x_s\overline{x_1x_2\ldots x_s} \\ &= x_1x_2\ldots x_s\bar{x}_s\ldots\bar{x}_1 \\ &= (-1)^s\alpha^s\beta(x_1,x_1)\ldots\beta(x_s,x_s) \end{aligned}$$

and if s is not even then $N(\tau)$ relative to $\alpha\beta$ may differ markedly from $N(\tau)$ relative to β . Hence, we restrict SN to SO(V) to assure that s is even.

Relating $SN : SO(V) \to R^*/R^{*2}$ to the group morphisms discussed at the end of Section (D), we have

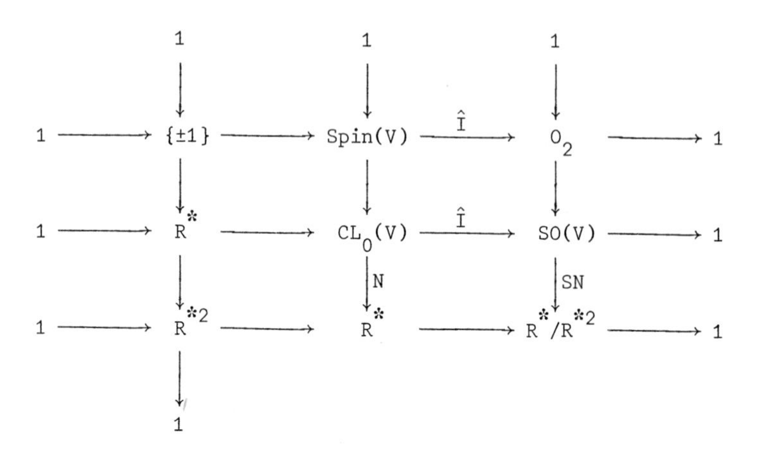

where O_2 is the <u>reduced orthogonal group</u>. It is perhaps worthwhile to note why $O_2 = \hat{I}(\mathrm{Spin}(V))$ is the kernel of SN . Certainly

$$O_2 \subset \mathrm{Kernel}(SN) .$$

On the other hand, suppose σ is in $SO(V)$ and $SN(\sigma) = 1$, i.e., σ is in the kernel of SN . Let x be in $CL_0(V)$ with $\hat{I}(x) = \sigma$. Then $N(x) = \alpha^2$ for α a unit in R^* . Then $\alpha^{-1}x$ is in the kernel of N , i.e., Spin(V) . But (a) $\hat{I}(\alpha^{-1}x) = \hat{I}(x) = \sigma$ and (b) $\hat{I}(\alpha^{-1}x)$ is in O_2 since $\alpha^{-1}x$ is in Spin(V) . Hence

$$O_2 = \mathrm{Kernel}(SN) .$$

We next show that, since the hyperbolic rank of V is ≥ 1 , both N and SN are surjective. We show this for SN -- the surjectivity of N is done similarly. Let ε be a unit of R . We need a σ in SO(V) with

$SN(\sigma) \equiv \varepsilon \mod R^{*2}$. The hypothesis of this section gives $V = H \perp W$ where $H = Ru \oplus Rv$ is a hyperbolic plane. Consider the isometry Φ_ε where $\Phi_\varepsilon(u) = \varepsilon u$, $\Phi_\varepsilon(v) = \varepsilon^{-1}v$ and $\Phi_\varepsilon|_W$ = identity . At the end of Section (C) we determined

$$\Phi_\varepsilon = \sigma_{u-v}\sigma_{u-\varepsilon v}$$

as a product of symmetries. Then Φ_ε is in $SO(V)$ and $I_{u-v}I_{u-\varepsilon v}|_V = \Phi_\varepsilon$. Hence $(u - v)(u - \varepsilon v)$ is in $CL_0(V)$ and

$$\begin{aligned} N((u - v)(u - \varepsilon v)) &= (-1)^2\beta(u - v,u - v)\beta(u - \varepsilon v,u - \varepsilon v) \\ &= 2^2\varepsilon \\ &\equiv \varepsilon \mod R^{*2} . \end{aligned}$$

Hence $SN(\Phi_\varepsilon) \equiv \varepsilon \mod R^{*2}$.

(V.28) THEOREM.[1] If the hyperbolic rank of V is ≥ 1 , the norm and spinor norm give the following commutative diagram having exact rows and columns.

[1] This result does not require 3 a unit. However, to show $\Omega(V) = O_2$ we need 3 a unit.

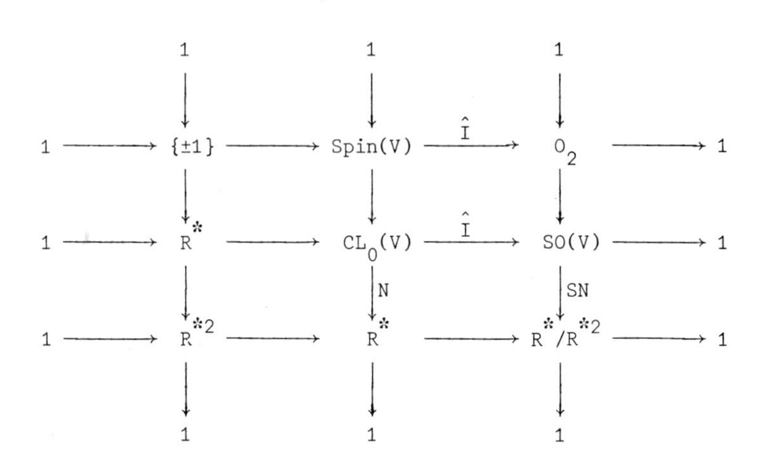

The above result is still not quite finished. Observe $\Omega(V) \subset \text{Kernel}(SN)$ and hence $\Omega(V) \subset O_2$. We show that, since the hyperbolic rank of V is ≥ 1 and 3 is a unit, $\Omega(V) = O_2$!

To verify the above remark, suppose τ is in the kernel of the spinor norm $SN : SO(V) \to R^*/R^{*2}$. Let $V = H \perp W$ where H is a hyperbolic plane. By the remarks and proofs in (III.D) there is a product w of unitary transvections (thus w is in $\Omega(V) = EO(V)$) satisfying

$$w\tau = \sigma$$

where $\sigma : H \to H$ and $\sigma|_W = \text{identity}$. Thus, the action of σ is confined to H and we may think of σ as an element of $SO(H)$.

Viewing σ as an element of SO(H) , by (V.4)(a),

$$\sigma = \sigma_x \sigma_y$$

for σ_x and σ_y symmetries. Since σ is in Ker(SN)

$$\begin{aligned} SN(\sigma) &= SN(\sigma_x \sigma_y) \\ &= \beta(x,x)\beta(y,y)R^{*2} \\ &= R^{*2} \end{aligned}$$

Hence, $\beta(x,x)\beta(y,y) = u^2$ for u a unit in R . Setting $\lambda = \frac{\beta(x,x)}{u}$ and $z = \lambda y$, we have a nonisotropic vector z in H with

$$\beta(x,x) = \beta(z,z) .$$

Further, by (V.1)(c),

$$\sigma_y = \sigma_{\lambda y} = \sigma_z .$$

By (V.3) there is a ρ in O(H) (which may be extended to O(V) by setting $\rho|_W$ = identity) which carries x to z , i.e., $\rho x = z$.

Hence,

$$\sigma = \sigma_x \sigma_y = \sigma_x \sigma_z = \sigma_x \sigma_{\rho x} = \sigma_x \rho \sigma_x \rho^{-1} = \sigma_x \rho \sigma_x^{-1} \rho^{-1}$$

and σ is in $\Omega(V)$.

We have proven the following theorem.

(V.29) THEOREM. If the hyperbolic rank of $V \geq 1$ and 3 is a unit, then the kernel of the spinor norm SN is precisely the commutator subgroup $\Omega(V)$.

The map SN may be extended naturally so that the commutator subgroups of various levels arise as its kernel. Let A be an ideal of R and define

$$SN_A = SN \times \lambda_A : SO(V) \to R^*/R^{*2} \times SO(V/AV)$$

by

$$SN_A : \sigma \to \langle SN(\sigma), \lambda_A\sigma \rangle .$$

We claim

$$\text{Kernel}(SN_A) = \Omega(V,A) .$$

Certainly, $\Omega(V,A) \subset \text{Kernel}(SN_A)$. Suppose τ is an element of the kernel of SN_A. Hence $SN(\tau) = R^{*2}$ and $\lambda_A(\tau) = I$. As in the above proof, letting $V = H \perp W$ where H is a hyperbolic plane, there is a w in $\Omega(V,A) = EO(V,A)$ with $w\tau = \sigma$ where $\sigma|_W = \text{identity}$. Since $\lambda_A w = \lambda_A \tau = 1$, we have $\lambda_A \sigma = 1$. A proof similar to the argument before (V.29) will represent σ as a commutator in $\Omega(V,A)$.

Thus, if hyperbolic rank of $V \geq 1$, then

$$1 \longrightarrow \Omega(V,A) \longrightarrow SO(V) \xrightarrow{SN_A} R^*/R^{*2} \times SO(V/AV)$$

is exact.

Additional results concerning SN_A may be found in Klingenberg [90].
The above proof technique will also give the following proposition (see (V.22)).

<u>(V.30)</u> <u>PROPOSITION</u>. Let A be an ideal of R, V have hyperbolic rank ≥ 1 and 3 be a unit. Then

$$1 \to \Omega(V,A) \to \Omega(V) \to \Omega(V/AV) \to 1$$

is exact.

Before terminating this section we remark on the role of "hyperbolic rank ≥ 1" in the study of $O(V)$.

If hyperbolic rank of $V \geq 1$ (2 is a unit — 3 likewise if necessary), one has a tower of groups

$$\begin{array}{c} O(V) \\ | \\ SO(V) \\ | \\ \Omega(V) \end{array} .$$

The determinant morphism shows $O(V)/SO(V) \simeq Z/Z2$ and the spinor norm shows $SO(V)/\Omega(V) \simeq R^*/R^{*2}$. Thus the bulk of the theory of $O(V)$ is contained in $\Omega(V)$. Here we find, since hyperbolic rank is ≥ 1, that $\Omega(V) = [\Omega(V),\Omega(V)]$ and $\Omega(V) = EO(V)$ is the Eichler group generated by unitary transvections. One speaks of the assumption of hyperbolic rank ≥ 1 as a "linearization" of the orthogonal group in the sense that $\Omega(V) = EO(V)$ in $O(V)$ plays a role similar to $E(V) = SL(V)$ in $GL(V)$. Intuitively, one would like to parallel the arguments for $E(V) = SL(V)$ in $GL(V)$ to obtain results for $EO(V)$ and $O(V)$.

(G) THE NORMAL SUBGROUPS OF O(V)

The determination of the normal subgroups of $O(V)$ is the central purpose of this section. In fact, slightly more is accomplished, we examine and classify the subgroups G of $O(V)$ which are normalized by $\Omega(V)$, i.e., the subgroups G of $O(V)$ where $\sigma\rho\sigma^{-1}$ is in G for all σ in $\Omega(V)$ and all ρ in G. The normal subgroups of $O(V)$ occur among the $\Omega(V)$-normal subgroups. An analogous situation occurred in (II) when the $SL(V)$-normal subgroups of $GL(V)$ were described. Actually in (II), we employed only conjugation by elementary transvections, i.e., elements of $E_B(V)$ — the group of elementary transvections relative to some fixed basis B. Since R was local, $E_B(V) = SL(V)$ and, consequently, we dealt with $SL(V)$. For more general commutative rings R, $E_B(V)$ sits properly within $SL(V)$. Further, distinct bases B often give rise to distinct $E_B(V)$.

We exploit the observation of the previous section that the bulk of the theory of $O(V)$ is determined by $\Omega(V)$ in a fashion analogous to $E(V)$ determining the structure of $GL(V)$. Indeed, this is the philosophical undercurrent in saying the presence of non-zero hyperbolic rank for V "linearizes" $O(V)$.

Of course, to utilize $\Omega(V)$ effectively, $\Omega(V)$ should behave in a reasonable fashion in $O(V)$. Consequently, the hypothesis of the previous section is continued, i.e., 2 is a unit of R, $\dim(V) \geq 3$, and V has hyperbolic rank ≥ 1. Thus, V splits as $V = H \perp W$ where $H = Ru \oplus Rv$ is a hyperbolic plane with $\beta(u,v) = 1$ and $\beta(u,u) = \beta(v,v) = 0$. When necessary other elements of R will also be assumed to be invertible. Often these additional assumptions may be replaced (if added care is shown in selecting units in R determined from units in the residue field $k = R/m$) by a weaker hypothesis on the cardinality of the residue class field k, e.g., replace " 3 is a unit" by " R/m is not the field of 3 elements." For expository motivations, we did not choose to present the arguments under these weaker hypotheses — details of the more general proofs can be found in James [87]. Our approach is modeled after the presentation of James [87] rather than that of Klingenberg [90]. The former exploits explicitly $\Omega(V)$ while the latter utilizes the spinor norm and $SO(V)$.

Let $\Omega(V)$ act on $O(V)$ as a transformation group under conjugation and examine the orbit of a single element ρ . We show that among the elements of this orbit there is an isometry which is a product of two unitary transvections, a suitable Φ_ε , and an element of $O(W)$. This result also extends to $O(V,A)$.

(V.31) THEOREM. Let A be an ideal of R and ρ be in $O(V,A)$. Then, for suitable ψ in $\Omega(V,A)$

$$\psi\rho\psi^{-1} = \sigma_1\sigma_2\Phi_\varepsilon\Theta$$

where

(a) σ_1 and σ_2 are unitary transvections of order contained in A ;

(b) $\varepsilon \equiv \pm 1 \mod A$;

(c) $\Theta|_H$ = identity . Hence Θ is in $O(W,A)$.

(Before we begin a proof it should be noted that σ_1 , σ_2 , Φ_ε or Θ may be trivial, e.g., ρ may be I . This will not affect the subsequent proofs.)

Proof. We first examine the case where $A \neq R$. Suppose

$$\rho(v) = \alpha u + \delta v + z$$

where α and δ are in R , z is in W and ρ is in $O(V,A)$.

Since $\lambda_A \rho = \pm I$, $\delta \equiv \pm 1$ modulo A and thus δ is a unit. Further, for the same reason, $\alpha \equiv 0$ modulo A and $\Pi_A z = 0$, i.e., $O(z) \subset A$. The general techniques are similar to (III.D).

The unitary transvection $\sigma_{u,\delta^{-1}z}$ is in $\Omega(V,A) = EO(V,A)$ and

$$\sigma_{u,\delta^{-1}z}(\rho(v)) = [\alpha + \frac{1}{2}\delta^{-1}\beta(z,z)]u + \delta v .$$

Then,

$$\Phi_\delta \sigma_{u,\delta^{-1}z}(\rho(v)) = [\delta\alpha + \frac{1}{2}\beta(z,z)]u + v .$$

Set $\rho_1 = \Phi_\delta \sigma_{u,\delta^{-1}z}\rho$. Since $\beta(v,v) = 0$, then $\beta(\rho_1(v),\rho_1(v)) = 0$ and, consequently, $\delta\alpha + \frac{1}{2}\beta(z,z) = 0$. Thus $\rho_1(v) = v$.

Next consider the action of ρ_1 on u . As above, in general,

$$\rho_1(u) = \gamma u + \mu v + w$$

for γ and μ in R and w in W . Note ρ_1 is in $O(V,A)$, thus $O(w) \subset A$. Since $1 = \beta(u,v) = \beta(\rho_1(u),\rho_1(v))$, we have $\gamma = 1$. Then

$$\sigma_{v,w}\rho_1(u) = u + \bar{\mu}v$$

and, using $\beta(u,u) = 0$, it is clear that $\bar{\mu} = 0$. Hence,

$$\sigma_{v,w}\Phi_\delta\sigma_{u,\delta^{-1}z}\rho = \Theta$$

where $\Theta(u) = u$ and $\Theta(v) = v$, i.e., Θ fixes H . Further, $O(w) \subset A$ implies $\sigma_{v,w}$ is in $\Omega(V,A)$. Then

$$\rho = \sigma_{u,-\delta^{-1}z} \Phi_{\delta^{-1}} \sigma_{v,-w} \Theta$$

$$= \sigma_{u,-\delta^{-1}z} \sigma_{v,-\delta w} \Phi_{\delta^{-1}} \Theta$$

by (III.16). This completes the case where $A \neq R$.

Suppose $A = R$. Let ρ be in $O(V)$ and suppose

$$\rho(v) = \alpha u + \delta v + z .$$

The proof will follow in the same manner as the above case $A \neq R$ provided δ is a unit. Thus, it is only necessary to show that ρ may be modified by conjugation so that δ is a unit. Let $\psi = \sigma_{v,-s}$ where s is in W . Clearly ψ is in $\Omega(V)$. Set $\bar{\rho} = \psi\rho\psi^{-1}$. Then

$$\begin{aligned} \bar{\rho}(v) &= \psi\rho\psi^{-1}(v) \\ &= \psi\rho(v) \\ &= \alpha u + [\delta - \beta(s,z) - \tfrac{1}{2}\alpha\beta(s,s)]v + (z + \alpha s) . \end{aligned}$$

First, if α is not initially a unit (along with δ), then $O(z) = R$ and we may select s in W with $\beta(s,z) = 1$. Thus $[\delta - \beta(s,z) - \frac{1}{2}\alpha\beta(s,s)]$ is a unit. Second, if α is a unit (δ a non-unit) and $O(z) \neq R$, then select s so that $\beta(s,s)$ is a unit. The resulting coefficient of v is then a unit. Third, if α is a unit (δ a non-unit) and $O(z) = R$, select s so that $\beta(s,z)$ is a unit. We are done unless $\frac{1}{2}\alpha\beta(s,s) \equiv -\beta(s,z)$ modulo m. If this case occurs, replace s by $2s$. Now proceed as in the case $A \neq R$ with ρ replaced by $\psi\rho\psi^{-1} = \bar{\rho}$.

In the above proof for the case $A = R$, $\bar{\bar{\rho}}$ is "essentially" $\bar{\bar{\rho}} = \psi\rho$ since ψ^{-1} has no effect on v. Thus, the proof provides the following decomposition of $O(V,A)$.

(V.32) COROLLARY. Let $V = H \perp W$ for H a hyperbolic plane. If A is an ideal of R, then

$$O(V,A) = SO(V,A)O(W,A) .$$

For the remainder of this section we fix the following setting:

Let G denote an $\Omega(V)$-normal subgroup of $O(V)$, i.e., G is a subgroup of $O(V)$ and

$$\sigma\rho\sigma^{-1} \text{ is in } G$$

for all σ in $\Omega(V)$ and ρ in G. For ρ in G, let C_ρ denote the orbit of ρ under conjugation by elements of $\Omega(V)$. Certainly, C_ρ is in G. Further, by (V.31), we may assume there is a ψ in $\Omega(V)$ with

$$\psi\rho\psi^{-1} = \sigma_{u,x}\sigma_{v,y}\Phi_\varepsilon\Theta$$

where $\Theta|_H$ = identity .

(V.33) LEMMA. (For the above setting) Assume 3 and 5 are also units in R . Then there are units ε_1 and ε_2 such that $\sigma_{u,\varepsilon_1 x}$ and $\sigma_{v,\varepsilon_2 y}$ belong to G .

Proof. Observe $\Phi_4 = [\Phi_2,\Delta]$ is in $\Omega(V)$. Thus the commutator

$$\begin{aligned}[\psi\rho\psi^{-1},\Phi_4] &= \psi\rho\psi^{-1}\Phi_4\psi\rho^{-1}\psi^{-1}\Phi_4^{-1} \\ &= \sigma_{u,x}\sigma_{v,y}\Phi_\varepsilon\Theta\Phi_4\Theta^{-1}\Phi_\varepsilon^{-1}\sigma_{v,-y}\sigma_{u,-x}\Phi_4^{-1} \\ &= \sigma_{u,x}\sigma_{v,y}\Phi_4\sigma_{v,-y}\sigma_{u,-x}\Phi_4^{-1} \\ &= \sigma_{u,x}\sigma_{v,y}\sigma_{v,-1/4y}\sigma_{u,-4x} \\ &= \sigma_{u,x}(\sigma_{v,-3/4y}\sigma_{u,-3x})\sigma_{u,x}^{-1}\end{aligned}$$

is in G . Conjugating by $\sigma_{u,x}$ shows that $\sigma_{v,-3/4y}\sigma_{u,-3x}$ is in G . It now suffices to show that if $\sigma_{u,x}\sigma_{v,y}$ lies in G , then there are units ε_1 and ε_2 such that $\sigma_{u,\varepsilon_1 x}$ and $\sigma_{v,\varepsilon_2 y}$ lie in G.

Observe since $\sigma_{u,x}\sigma_{v,y}$ is in G then

$$\sigma_{v,y}\sigma_{u,x} = \sigma_{u,x}^{-1}(\sigma_{u,x}\sigma_{v,y})\sigma_{u,x}$$

is in G . Thus

$$\sigma_{u,2x}\sigma_{v,2y} = \sigma_{v,y}^{-1}(\sigma_{v,y}\sigma_{u,x})(\sigma_{u,x}\sigma_{v,y})\sigma_{v,y}$$

is in G .

Repeating the argument of the previous paragraph shows that

$$\sigma_{u,4x}\sigma_{v,4y}$$

is in G . Then

$$\sigma_{u,16x}\sigma_{v,y} = \Phi_4\sigma_{u,4x}\sigma_{v,4y}\Phi_4^{-1}$$

is in G since as noted above Φ_4 is in $\Omega(V)$. Then

$$\sigma_{u,15x}\sigma_{v,y}\sigma_{u,x} = \sigma_{u,x}^{-1}(\sigma_{u,16x}\sigma_{v,y})\sigma_{u,x}$$

is in G and since $\sigma_{v,y}\sigma_{u,x}$ is in G , we conclude $\sigma_{u,15x}$ is in G . Hence, there is a unit ε_1 such that $\sigma_{u,\varepsilon_1 x}$ is in G . An analogous argument shows there is a unit ε_2 such that $\sigma_{v,\varepsilon_2 y}$ is in G . This completes the proof.

Observe if one is interested only in the normal subgroups of $O(V)$, rather than the $\Omega(V)$-normal subgroups, then in the above proof the conjugation by Φ_4 (which lies in $\Omega(V)$) may be replaced by conjugation by Φ_2 (which is in $O(V)$). Then, with minor modifications, the argument will carry through under only the hypothesis that 2 and 3 are units, i.e., we may omit the assumption that 5 is a unit. Also, as noted earlier, the hypothesis on the existence of units may be replaced by a weaker hypothesis on the size of the residue class field (see [87]).

Having obtained $\sigma_{u,x}$ in G, we next show that suitable conjugation gives

$$\Omega(V,\mathcal{O}(x)) \leq G .$$

That is, the orbit of $\sigma_{u,x}$ when $\Omega(V)$ acts on $O(V)$ via conjugation is $\Omega(V,\mathcal{O}(x))$.

<u>(V.34)</u> <u>THEOREM</u>. (For the above setting) Assume $\dim(V) \neq 4$. If $\sigma_{u,x}$ is in G, then

$$\Omega(V,O(x)) \leq G .$$

<u>Proof</u>. The proof will follow from a series of steps.

(a) If η is in R, we show that η may be written as a sum or difference of squares of units.

Suppose η is in m. Then $1 + \frac{\eta}{4}$ and $1 - \frac{\eta}{4}$ are units and

$$\eta = (1 + \tfrac{\eta}{4})^2 - (1 - \tfrac{\eta}{4})^2 .$$

Suppose η is not in m. Then η is a unit. Since $\eta = (\frac{\eta}{2} + 1) + (\frac{\eta}{2} - 1)$ either $\frac{\eta}{2} + 1$ or $\frac{\eta}{2} - 1$ is a unit. Assume

$\frac{\eta}{2} + 1$ is a unit. Then

$$\eta = (1 + \tfrac{\eta}{2})^2 - (\tfrac{\eta}{2})^2 - 1 .$$

The case $\frac{\eta}{2} - 1$ is analogous.

(b) If $\sigma_{u,x}$ is in G, then $\sigma_{u,\eta x}$ is in G for all η in R.

Observe, if δ is a unit, then Φ_{δ^2} is in $\Omega(V)$ since $\Phi_{\delta^2} = [\Phi_\delta, \Delta]$. Then, if $\sigma_{u,x}$ is in G, we have

$$\sigma_{u,\delta^2 x} = \Phi_{\delta^2} \sigma_{u,x} \Phi^{-1}_{\delta^2}$$

in G. Suppose, by (a), η in R may be written as $\eta = \delta_1^2 - \delta_2^2 - 1$ for δ_1, δ_2 units in R. Then,

$$\sigma_{u,\eta x} = \sigma_{u,\delta_1^2 x} \sigma^{-1}_{u,\delta_2^2 x} \sigma^{-1}_{u,x}$$

is in G. The other cases in (a) are handled similarly.

The above argument shows that $\Omega(V,O(x)) \leq G$ in the case $\dim(V) = 3$. Thus, for the remainder of the proof we assume $\dim(V) \geq 5$ and, consequently, $V = H \perp W$ where $\dim(W) = m \geq 3$.

(c) We next manufacture some elements in $\Omega(V)$.

As a final remark to Section (V.C), we noted the symmetry σ_z

for z nonisotropic in W could be expressed as

$$\sigma_z = \Delta\Phi_{-\frac{1}{2}\beta(z,z)}\sigma_{v,-z}\sigma_{u,-\frac{2z}{\beta(z,z)}}\sigma_{v,-z}$$

in terms of the unitary transvections, Δ and Φ_ε. Recall, from (III.16), $\Delta^2 = I$ and $\Delta\Phi_\varepsilon\Delta = \Phi_{\varepsilon^{-1}}$. Hence, letting $\varepsilon = -\frac{1}{2}\beta(z,z)$,

$$\Phi_\varepsilon^{-1}\Delta\sigma_z = \sigma_{v,-z}\sigma_{u,\varepsilon^{-1}z}\sigma_{v,-z}$$

is in $\Omega(V)$.

Consequently, for y and z nonisotropic in W the isometry ($\alpha = -\frac{1}{2}\beta(y,y)$)

$$\begin{aligned}
&\Phi_\varepsilon^{-1}\Delta\sigma_z\Phi_\alpha^{-1}\Delta\sigma_y \\
&\quad = \Phi_{\varepsilon^{-1}}\Phi_\alpha\Delta^2\sigma_z\sigma_y && \text{(by (III.16)(b))} \\
&\quad = \Phi_{\varepsilon^{-1}}\Phi_\alpha\sigma_z\sigma_y && \text{(since } \Delta^2 = I \text{)}
\end{aligned}$$

is in $\Omega(V)$.

By (V.1) $\sigma_z = \sigma_{\lambda z}$ for λ a unit in R. Thus, in $\Phi_{\varepsilon^{-1}}\Phi_\alpha\sigma_z\sigma_y$ replace z by $\beta(z,z)^{-1}z$ and y by $\frac{1}{2}y$. The factor $\sigma_z\sigma_y$ is unchanged. However, $\Phi_{\varepsilon^{-1}}\Phi_\alpha = \Phi_{\varepsilon^{-1}\alpha}$ becomes

$$\Phi_{\frac{\beta(z,z)\beta(y,y)}{4}}$$

under these replacements.

Therefore,

$$\Phi_{\frac{\beta(z,z)\beta(y,y)}{4}}\sigma_z\sigma_y$$

is in $\Omega(V)$.

(d) Suppose, further that y and z are nonisotropic in W and satisfy $\beta(y,z) = 1$. A direct computation gives (see (V.1)(d))

$$(\sigma_z\sigma_y - \sigma_y\sigma_z)(x) = \frac{4}{\beta(z,z)\beta(y,y)}[\beta(y,x)z - \beta(z,x)y] .$$

Hence, if $\varepsilon = \frac{1}{4}\beta(z,z)\beta(y,y)$ (as above), then

$$\varepsilon(\sigma_z\sigma_y - \sigma_y\sigma_z)(x) = \beta(y,x)z - \beta(z,x)y .$$

(e) We next combine (d) and (c) above and apply them to the unitary transvection $\sigma_{u,x}$ in G . Since G is $\Omega(V)$-normal and $\Phi_\varepsilon\sigma_z\sigma_y$ is in $\Omega(V)$ where z , y and ε are as given in (d), we have

$$(\Phi_\varepsilon\sigma_z\sigma_y)\sigma_{u,x}(\Phi_\varepsilon\sigma_z\sigma_y)^{-1} = \sigma_{u,\varepsilon\sigma_z\sigma_y x}$$

in G by (III.16). Likewise $\sigma_{u,\varepsilon\sigma_y\sigma_z x}$ is in G . Thus,

$$\begin{aligned}\sigma_{u,\varepsilon\sigma_z\sigma_y x}\sigma^{-1}_{u,\varepsilon\sigma_y\sigma_z x} &= \sigma_{u,\varepsilon\sigma_z\sigma_y x}\sigma_{u,-\varepsilon\sigma_y\sigma_z x}\\ &= \sigma_{u,\varepsilon\sigma_z\sigma_y x-\varepsilon\sigma_y\sigma_z x}\\ &= \sigma_{u,\beta(y,x)z-\beta(z,x)y}\end{aligned}$$

is in G by (III.16).

We will now complete the proof by wisely choosing x, y and z.

(f) Suppose $\sigma_{u,x}$ is in G. Let W have an orthogonal basis $\{x_1,x_2,\ldots,x_m\}$ where $m = n - 2 \geq 3$. Then $x = \alpha_1 x_1 + \cdots + \alpha_m x_m$ and

$$O(x) = (\alpha_1,\ldots,\alpha_m) .$$

We want to show $\Omega(V,O(x)) \subset G$. Since $\Omega(V,O(x)) = EO(V,O(x))$, we need to show $\sigma_{u,\bar{x}}$ is in G where $O(\bar{x})$ is in $(\alpha_1,\ldots,\alpha_m)$. (A similar argument employing $\sigma_{v,x}$ will place $\sigma_{v,\bar{x}}$ in G.) If $\bar{x} = \delta_1 x_1 + \cdots + \delta_m x_m$ then δ_i is in $O(x)$ and, consequently, $\delta_i = \sum_j \mu_{ij}\alpha_j$. Hence, $\bar{x} = \sum_{ij}\sum \mu_{ij}\alpha_j x_i$.

Thus, if we are to show $\sigma_{u,\bar{x}}$ is in G, we need first to show $\sigma_{u,\alpha_j x_i}$ is in G for each i and j. Then, by (b) of this proof, $\sigma_{u,\mu_{ij}\alpha_j x_i}$ will be in G for all μ_{ij} in R.

In turn, by (III.16),

$$\sigma_{u,\bar{x}} = \sigma_{u,\sum \mu_{ij}\alpha_j x_i}$$

$$= \Pi \sigma_{u,\mu_{ij}\alpha_j x_i}$$

is in G.

We use part (e) above, i.e., if y and z are nonisotropic in W with $\beta(y,z) = 1$ then

(*) $$\sigma_{u,\beta(y,x)z-\beta(z,x)y}$$

is in G whenever $\sigma_{u,x}$ is in G.

Let $\varepsilon_i = \beta(x_i,x_i)$ for $1 \le i \le n$. Let $y_i = \frac{1}{\varepsilon_i} x_i$. Then $\beta(x_i,y_i) = 1$ for $1 \le i \le n$.

Let $z = x_i + \eta x_j$ and $y = y_i$ where $i \neq j$ and η is chosen so that $\beta(z,z)$ is a unit. By substituting in (*)

$$\sigma_{u,\beta(y,x)z-\beta(z,x)y} = \sigma_{u,\eta(\alpha_i x_j - \alpha_j \varepsilon_j y_i)} .$$

Since η may be chosen arbitrarily except for at most two classes modulo the maximal ideal m, we may take η to be a unit. (Note, 1, 2 and 3 are units in R/m.) Thus, by part (b),

$$\sigma_{u,\bar{\eta}(\alpha_i x_j - \alpha_j \varepsilon_j y_i)}$$

is in G for all $\bar{\eta}$ in R.

Let $1 \le i,j,k \le m$ be distinct. (Recall $m \ge 3$.) In (*) set

$$x = \bar{\eta}(\alpha_i x_j - \alpha_j \varepsilon_j y_i) ,$$
$$z = x_k ,$$
$$y = y_k + \delta y_j ,$$

where δ is a unit and selected so that $\beta(z,y) = 1$. Then,

$$\sigma_{u,\beta(y,x)z-\beta(z,x)y} = \sigma_{u,\delta\bar{\eta}\alpha_i x_k}$$

is in G where $i \neq k$. Since $\delta\bar{\eta}$ may be chosen to be a unit, by part (b),

$$\sigma_{u,\alpha_i x_k}$$

is in G for all i and k with $i \neq k$.

It remains to show $\sigma_{u,\alpha_i x_i}$ is in G for $1 \leq i \leq m$.

In (*) set

$$x = \alpha_i x_k ,$$

$$z = x_k + \bar{\bar{\eta}} x_i ,$$

$$y = \frac{x_k}{\varepsilon_k}$$

where $\bar{\bar{\eta}}$ is a unit with z nonisotropic and $i \neq k$. Then

$$\sigma_{u,\beta(y,x)z-\beta(z,x)y} = \sigma_{u,\bar{\bar{\eta}}\alpha_i x_i} .$$

By (b), $\sigma_{u,\alpha_i x_i}$ is in G.

This completes the proof of (V.34). An analogous statement and proof will apply for $\sigma_{v,x}$ in G.

We now can prove the main theorem of this section. We state this result with its complete hypothesis.

(V.35) THEOREM. Let R be a local ring with 2, 3 and 5 being units. Let $\dim(V) = 3$ or $\dim(V) \geq 5$ and assume V has hyperbolic rank ≥ 1. Suppose G is a subgroup of $O(V)$ normalized by $\Omega(V)$ and let A be an ideal which is maximal with respect to $\Omega(V,A) \leq G$. Then

$$\Omega(V,A) \leq G \leq O(V,A) .$$

Proof. Observe $\Omega(V,0) = I$ is always in G. Thus, there are ideals B with $\Omega(V,B) \leq G$.

If $\{B_\lambda\}$, $\lambda \in \Lambda$, is a family of ideals satisfying $\Omega(V,B_\lambda) \leq G$ then, since $EO(V,B_\lambda) = \Omega(V,B_\lambda)$, it is easy to see

$$\Omega(V,\sum_\lambda B_\lambda) \leq G .$$

Thus, the A in the theorem is unique and contains every ideal B with $\Omega(V,B) \leq G$.

Let ϕ be in G. By (V.31) there is a ψ in $\Omega(V)$ with

$$\psi\phi\psi^{-1} = \sigma_{u,x}\sigma_{v,y}\Phi_\varepsilon\Theta .$$

By (V.33) and (V.34), $\sigma_{u,x}$ and $\sigma_{v,y}$ are in G . Thus $\Phi_\varepsilon\Theta$ is in G .

For arbitrary z in W , using (III.16), we have, since $\Theta|_H$ = identity ,

$$[\sigma_{u,-z},\Phi_\varepsilon\Theta] = \sigma_{u,\varepsilon\Theta(z)-z}$$

in G . By (V.34),

$$\Omega(V,0(\varepsilon\Theta(z) - z)) \leq G$$

and, consequently, by the above remark

$$0(\varepsilon\Theta(z) - z) \subset A \ .$$

Therefore

$$\varepsilon\Theta(z) \equiv z \mod A \ .$$

Select z in W with z nonisotropic, i.e., $\beta(z,z)$ is a unit. Then

$$\begin{aligned} \varepsilon^2\beta(\Theta(z),\Theta(z)) &= \beta(\varepsilon\Theta(z),\varepsilon\Theta(z)) \\ &\equiv \beta(z,z) \mod A \ . \end{aligned}$$

Since $\beta(z,z)$ is a unit, $\varepsilon^2 \equiv 1 \mod A$ and thus

$$\Theta(z) \equiv \varepsilon z \mod A$$

for all z in W .

If we apply the above to ϕ ,

$$\phi(x) \equiv \varepsilon x \mod A$$

for all x in V . But 2 is a unit. Hence $\varepsilon \equiv \pm 1 \mod A$ and thus $\phi \equiv \pm I \mod A$, i.e., ϕ is in $O(V,A)$. This finishes the proof.

We recall the remark after (V.33). Thus, if we ask G to be normal rather than $\Omega(V)$-normal then the assumption that 5 is a unit may be omitted. For other comments and remarks see James [87].

We conclude this section with several useful observations.

Remark (I). Let G be a subgroup of $O(V)$. The order of G , denoted $o(G)$, is the smallest ideal A satisfying

$$\lambda_A G \leq \text{Center } O(V/AV) .$$

Thus, since $\dim(V) \geq 3$, $o(G)$ is the smallest ideal A with $\lambda_A \sigma = \pm I$ for all σ in G .

Suppose G is $\Omega(V)$-normal and $o(G) = A$. Under the hypothesis of (V.35), there is an ideal B with

$$\Omega(V,B) \leq G \leq O(V,B) .$$

Thus, since $G \leq O(V,B)$, we have A contained in B . On the other hand, (V.19) gives $\Omega(V,B) = EO(V,B)$ and clearly the generators of $EO(V,B)$ indicate $o(EO(V,B)) = B$. Hence $o(\Omega(V,B)) = B$ and since $\Omega(V,B) \leq G$, B is contained in A . Hence $A = B$. Thus, if G is an $\Omega(V)$-normal subgroup of $O(V)$ and $o(G) = A$, then

$$\Omega(V,A) \leq G \leq (V,A) .$$

Remark (II). Suppose G is any subgroup of $O(V)$ satisfying

$$\Omega(V,A) \leq G \leq O(V,A)$$

for an ideal A of R.

Let σ be in $O(V)$ and τ be in G. Then $\sigma\tau\sigma^{-1}\tau^{-1}$ is in $\Omega(V,A)$ since τ is in $O(V,A)$. Hence $\sigma\tau\sigma^{-1}\tau^{-1} = \rho$ for some ρ in G since $\Omega(V,A)$ is in G. That is, $\sigma\tau\sigma^{-1} = \rho\tau$ is in G and G is normal in $O(V)$. Therefore, if G is any subgroup of $O(V)$ satisfying

$$\Omega(V,A) \leq G \leq O(V,A)$$

for an ideal A, then G is normal (hence $\Omega(V)$-normal) in $O(V)$.

(H) INVOLUTIONS

This is the first of the sections dealing with the automorphisms of the orthogonal group. Our purpose in this section is to show an automorphism of $O(V)$ will carry extremal involutions to extremal involutions.

Let σ be an involution in $O(V)$, i.e., $\sigma^2 = I$. As in (II.E), σ decomposes the space V into

$$V = P(\sigma) \oplus N(\sigma)$$

where $P(\sigma) = \{x \text{ in } V \mid \sigma x = x\}$ is the positive space of σ and

$N(\sigma) = \{x \text{ in } V \mid \sigma x = -x\}$ is the <u>negative space</u> of σ. The subspaces $P(\sigma)$ and $N(\sigma)$ are called the <u>proper spaces</u> of σ. Further, since σ is in $O(V)$, the splitting of V by the proper spaces of σ is an orthogonal sum, i.e.,

$$V = P(\sigma) \perp N(\sigma) .$$

If $P(\sigma)$ is either a line or a hyperplane, then σ is called an <u>extremal</u> involution. Observe, if $P(\sigma)$ is a hyperplane, then σ is a symmetry. Let Ext_β denote the set of extremal involutions in $O(V)$. Recall $\beta : V \times V \to R$ is the symmetric inner product on V.

Suppose σ is an extremal involution in $O(V)$. Then the proper spaces of σ are a line and a hyperplane. Let

$$L(\sigma) = \text{proper line of } \sigma \text{, and}$$
$$H(\sigma) = \text{proper hyperplane of } \sigma .$$

Then $L(\sigma)$ and $H(\sigma)$ determine σ up to ± 1.

We recall some facts from (II.E) or (IV.E). Suppose σ is an involution in $O(V)$ with proper spaces P and N. If ρ is in $O(V)$, then $\rho\sigma\rho^{-1}$ is an involution with proper spaces ρP and ρN. Further, $\rho\sigma = \sigma\rho$ if and only if $\rho P \subseteq P$ and $\rho N \subseteq N$. In particular, suppose ρ and σ are extremal involutions. Then $\rho\sigma = \sigma\rho$ and $\sigma \neq \pm\rho$ if and only if $L(\sigma) \subset H(\rho)$ and $L(\rho) \subset H(\sigma)$.

Suppose $\alpha : V \to V$ is an R-morphism in $\mathrm{End}(V)$. Employing the inner product β, we define an R-morphism

$$\alpha^* : V \to V$$

by

$$\beta(\alpha^* v,w) = \beta(v,\alpha w) .$$

We call α^* the adjoint of α. Note this is consistent with (I) and (II) and, for convenience, we write the action of the adjoint on the left of V. Observe α determines α^* uniquely. If $\alpha^* = \alpha$ then α is called self-adjoint. Finally, an R-morphism $\Pi : V \to V$ is a projection or idempotent if $\Pi^2 = \Pi$.

Suppose Π is a self-adjoint projection. We claim

$$V = \mathrm{Im}(\Pi) \perp \mathrm{Im}(\Pi)^{\perp} .$$

To show this suppose first that x is in $\mathrm{Im}(\Pi) \cap \mathrm{Im}(\Pi)^{\perp}$. Then, since x is in $\mathrm{Im}(\Pi)$, $\Pi x = x$ and thus for each y in V,

$$\beta(y,x) = \beta(y,\Pi x) = \beta(\Pi y,x) .$$

But the last inner product is 0 since x is in $\mathrm{Im}(\Pi)^{\perp}$. Hence $\beta(y,x) = 0$ for all y in V and, consequently, $x = 0$. Thus, $\mathrm{Im}(\Pi) \cap \mathrm{Im}(\Pi)^{\perp} = 0$. Second, let y be in V and note

$$y = \Pi y + (y - \Pi y) .$$

Certainly Πy is in $\mathrm{Im}(\Pi)$. To show $y - \Pi y$ is in $\mathrm{Im}(\Pi)^{\perp}$ consider any z in V . Then

$$\begin{aligned}\beta(\Pi z, y - \Pi y) &= \beta(\Pi z, y) - \beta(\Pi z, \Pi y) \\ &= \beta(\Pi^2 z, y) - \beta(\Pi z, \Pi y) \\ &= \beta(\Pi z, \Pi y) - \beta(\Pi z, \Pi y) \\ &= 0\end{aligned}$$

Hence, $y - \Pi y$ is in $\mathrm{Im}(\Pi)^{\perp}$. The following lemma is now straightforward.

<u>(V.36)</u> <u>LEMMA</u>. Let Π be a self-adjoint projection. Then,

(a) $V = \mathrm{Im}(\Pi) \perp \mathrm{Im}(\Pi)^{\perp}$.

(b) $\mathrm{Im}(I - \Pi) = \mathrm{Im}(\Pi)^{\perp}$.

(c) If $V = W \perp W^{\perp}$ for a subspace W , then the projection $\Pi_W : V \to W$ is self-adjoint.

If σ in $O(V)$ is an involution, then $\rho = I - \sigma$ is an R-morphism of V . We will be interested in when such an R-morphism commutes with a self-adjoint projection.

<u>(V.37)</u> <u>LEMMA</u>. Let $\rho : V \to V$ be an R-morphism and Π be a self-adjoint projection. If $\rho\Pi = \Pi\rho = \rho$, then

(a) $\rho(I - \Pi) = 0$,

(b) $\mathrm{Im}(I - \Pi) = \mathrm{Im}(\Pi)^{\perp} \subset \mathrm{Ker}(\rho)$,

(c) $\mathrm{Im}(\rho) \subset \mathrm{Im}(\Pi)$.

Letting $\rho = I - \sigma$ for σ an involution in $O(V)$, we have the following lemma.

<u>(V.38)</u> <u>LEMMA</u>. Let σ in $O(V)$ be an involution. If $\rho = I - \sigma$, then

(a) $\mathrm{Im}(\rho) = N(\sigma)$ (negative space of σ).

(b) $\mathrm{Ker}(\rho) = P(\sigma)$ (positive space of σ).

<u>Proof</u>. For (a), suppose $y = \rho x$, i.e., y is in $\mathrm{Im}(\rho)$. Then $\sigma y = \sigma(\rho x) = \sigma(I - \sigma)x = (\sigma - \sigma^2)x = (\sigma - I)x = -(I - \sigma)x = -\rho x = -y$. On the other hand, if y is in $N(\sigma)$ then $(\frac{1}{2})y$ is in $N(\sigma)$. Thus, $\rho((\frac{1}{2})y) = (I - \sigma)((\frac{1}{2})y) = (\frac{1}{2})y + (\frac{1}{2})y = y$ and y is in $\mathrm{Im}(\rho)$.

For (b), observe the following are equivalent: x in $\mathrm{Ker}(\rho)$; $\rho(x) = 0$; $(I - \sigma)(x) = 0$; $x = \sigma(x)$; x in $P(\sigma)$.

The above lemmas indicate that if σ in $O(V)$ is an involution and $V = W \perp W^{\perp}$ is an orthogonal splitting of V, then, for $\Pi = \Pi_W : V \to W$ and $\rho = I - \sigma$, we have the following:

If $\Pi = \rho\Pi = \Pi\rho$ or, equivalently, $\sigma\Pi = \Pi\sigma = 0$, then

$$\mathrm{Im}(\rho) = N(\sigma) \subset \mathrm{Im}(\Pi) = W$$

and

$$\mathrm{Ker}(\rho) = P(\sigma) \supset \mathrm{Im}(\Pi)^{\perp} = W^{\perp}.$$

<u>(V.39)</u> <u>THEOREM</u>. Let Π be a self-adjoint projection. Let S denote the set of all extremal involutions σ with $\sigma\Pi = \Pi\sigma$. Suppose ρ is an involution in $O(V)$ which commutes with all σ in S . Then

$$\rho\Pi = \pm\Pi \quad .$$

<u>Proof</u>. Let $Im(\Pi) = W$. Since W splits V , W has an orthogonal basis, say $\{x_1,\ldots,x_t\}$, $t \leq n$. Then, the symmetries $\sigma_{x_1},\ldots,\sigma_{x_t}$ are extremal involutions commuting with Π .

Thus $\rho\sigma_{x_i} = \sigma_{x_i}\rho$ and, consequently, $\rho N(\sigma_{x_i}) = N(\sigma_{x_i})$, i.e., $\rho Rx_i = Rx_i$ for $1 \leq i \leq t$. Hence $\rho W \subseteq W$.

Let $x = x_i$ for some i , $1 \leq i \leq t$. Then $\rho x = \delta x$ for some δ in R . Since $\rho^2 = I$, we have $\delta^2 = 1$ and $\delta = \pm 1$. Therefore $\rho x_i = \pm x_i$ for each i . We show next the sign $(\pm)$ is the same for each i . If $t = 1$, this is trivial. Suppose $t \geq 2$. Let x and y be distinct in $\{x_1,\ldots,x_t\}$. Then, there is a unit α in R with $x + \alpha y$ nonisotropic (indeed, α may be chosen to be 1 or 2). Then $\sigma_{x+\alpha y}$ is in S and thus $\rho R(x + \alpha y) = R(x + \alpha y)$. Let

$$\rho x = \delta x \ ,$$
$$\rho y = \bar{\delta} y \ , \text{ and}$$
$$\rho(x + \alpha y) = \hat{\delta}(x + \alpha y)$$

where δ , $\bar{\delta}$ and $\hat{\delta}$ are in $\{\pm 1\}$.

Equating these results gives

$$\delta x + \alpha\bar{\delta} y = \hat{\delta}(x + \alpha y) .$$

Hence $\delta = \bar{\delta} = \hat{\delta}$. Thus, for all x in W either $\rho x = x$ or $\rho x = -x$.

Let X and Y be subsets of $O(V)$. The <u>centralizer</u> of Y by X in $O(V)$ is

$$C_X(Y) = \{\sigma \text{ in } X \mid \sigma\rho = \rho\sigma \text{ for all } \rho \text{ in } Y\} .$$

Observe if Λ is an automorphism of $O(V)$, then

$$\Lambda C_X(Y) = C_{\Lambda X}(\Lambda Y) .$$

Let Inv denote the set of involutions of $O(V)$. Then denote $C_{\mathrm{Inv}}(Y)$ by $C(Y)$ and define the <u>double centralizer</u> $C^2(Y)$ to be $C(C(Y))$. Suppose σ and ρ are involutions in $O(V)$. Define

$$\#(\sigma,\rho) = \text{the number of distinct involutions in } C^2(\{\sigma,\rho\}) .$$

Then, define

$$\max(\sigma) = \max\{\#(\sigma,\rho) \mid \rho \text{ is an involution and } \rho\sigma = \sigma\rho\} .$$

We use the above to give a group theoretic characterization of extremal involutions in $O(V)$.

(V.40) THEOREM. An involution σ in $O(V)$ is extremal if and only if $\max(\sigma) = 8$.

Proof. Recall from (II.E) if X is any set of pairwise commuting involutions in $O(V)$, then V splits as

$$V = L_1 \perp L_2 \perp \cdots \perp L_n$$

where (1) the L_i are lines,

(2) for each σ in X , $\sigma|_{L_i} = \pm I$.

Hence, the proper spaces of each σ are composed of orthogonal sums of elements in $\{L_1, \ldots, L_n\}$.

Let σ and ρ be commuting involutions in $O(V)$. By the above their proper spaces are orthogonal sums of suitable lines L_i , $1 \le i \le n$. Hence,

$$U_1 = P(\sigma) \cap P(\rho) ,$$

$$U_2 = P(\sigma) \cap N(\rho) ,$$

$$U_3 = N(\sigma) \cap P(\rho) , \text{ and}$$

$$U_4 = N(\sigma) \cap N(\rho)$$

are subspaces. Let $\Pi_i : V \to U_i$ be the projection, $1 \le i \le 4$.

Suppose α is an extremal involution in $C(\{\sigma,\rho\})$. Then, $\alpha\rho = \rho\alpha$ and $\alpha\sigma = \sigma\alpha$. Thus, α induces a further splitting of the proper spaces of ρ and σ and, in particular, $\Pi_i\alpha = \alpha\Pi_i$ for $1 \le i \le 4$. Conversely, an extremal involution α in $O(V)$ satisfying $\Pi_i\alpha = \alpha\Pi_i$ commutes with σ and ρ.

Then, if τ is any involution in the double centralizer $C^2(\{\sigma,\rho\})$, τ must commute with all extremal involutions α in $C(\{\sigma,\rho\})$. Since the Π_i are self-adjoint, apply (V.39), and conclude $\tau\Pi_i = \pm\Pi_i$ for $1 \le i \le 4$. However, $I = \Pi_1 + \Pi_2 + \Pi_3 + \Pi_4$ so

$$\begin{aligned}\tau = \tau I &= \tau\Pi_1 + \tau\Pi_2 + \tau\Pi_3 + \tau\Pi_4 \\ &= \delta_1\Pi_1 + \delta_2\Pi_2 + \delta_3\Pi_3 + \delta_4\Pi_4\end{aligned}$$

where $\delta_i = \pm 1$.

Suppose σ is a non-trivial involution in $O(V)$. Then $\dim(P(\sigma)) \ge 1$ and $\dim(N(\sigma)) \ge 1$.

Suppose

$$\begin{aligned}P(\sigma) &= Re_1 \perp \cdots \perp Re_t\,, \\ N(\sigma) &= Re_{t+1} \perp \cdots \perp Re_n\end{aligned}$$

where the $\{e_i\}$ forms an orthogonal basis. Let ρ satisfy $N(\rho) = Re_1 \perp Re_n$ and $P(\rho) = Re_2 \perp \cdots \perp Re_{n-1}$. Then, the above subspaces are

$$U_1 = Re_2 \perp \cdots \perp Re_t ,$$
$$U_2 = Re_1 ,$$
$$U_3 = Re_{t+1} \perp \cdots \perp Re_{n-1} \quad \text{and}$$
$$U_4 = Re_n$$

Then, at most only one U_i is 0. This occurs if and only if $t = 1$ or $t = n - 1$, i.e., if and only if σ is extremal. Thus, if σ is extremal, there are 8 choices for

$$\tau = (\pm)\Pi_1 + (\pm)\Pi_2 + (\pm)\Pi_3 + (\pm)\Pi_4$$

in $C^2(\{\sigma,\rho\})$ since one of the Π_i is 0. If σ is not extremal there are 16 choices and $\max(\sigma) = 16$.

<u>(V.41)</u> <u>THEOREM</u>. Let Ext_β denote the extremal involutions in $O(V)$. Let $\Lambda : O(V) \to O(V)$ be a group automorphism. Then

$$\Lambda(Ext_\beta) = Ext_\beta .$$

<u>Proof</u>. Let σ be an extremal involution. Then

$$|C^2(\{\sigma,\rho\})| = |C^2(\{\Lambda\sigma,\Lambda\rho\})|$$

for ρ an involution commuting with σ. Hence,

$$\max(\Lambda\sigma) = \max(\sigma) = 8$$

and $\Lambda\sigma$ is an extremal involution.

(I) THE STANDARD AUTOMORPHISMS

This section, as did (IV.G) and (II.F), discusses the standard automorphisms of $O(V)$. As expected, they fall into two classes: (1) inner-automorphisms Φ_g induced by semi-linear isomorphisms $g : V \to V$ and (2) radial automorphisms P_χ induced by group morphisms $\chi : O(V) \to \text{Center}(O(V))$.

Suppose $g : V \to V$ is a semi-linear isomorphism of V with associated ring automorphism μ . Then

$$\Phi_g(\sigma) = g\sigma g^{-1} \qquad (\ \sigma \text{ in } GL(V)\ )$$

is a group automorphism of $GL(V)$ and satisfies

(a) $\Phi_{g_1} \circ \Phi_{g_2} = \Phi_{g_1 g_2}$,

(b) $\Phi_g^{-1} = \Phi_{g^{-1}}$.

If G is a subgroup of $GL(V)$, for example, $G = O(V)$, then Φ_g is <u>on</u> G if

$$\Phi_g(G) = G .$$

Thus, Φ_g is on G if $\Phi_g|_G$ is an automorphism of G . Equivalently, Φ_g induces an automorphism of G . We will denote $\Phi_g|_G$ as well as the original automorphism by Φ_g .

As in (IV.G), Φ_g will be on $O(V)$ if g behaves "properly" with regard to the symmetric inner product β . The semi-linear isomorphism g is said to preserve orthogonality if for all unimodular x and y in V we have $\beta(x,y) = 0$ implies $\beta(gx,gy) = 0$. We will show in (V.44) that Φ_g is on $O(V)$ if and only if g preserves orthogonality.

(V.42) PROPOSITION. The semi-linear isomorphism $g : V \to V$ preserves orthogonality if and only if there is a unit α in R with

$$\beta(gx,gy) = \alpha(\beta(x,y))^{\mu}$$

for all x and y in V where $\mu : x \to x^{\mu}$ is the ring isomorphism associated with g .

Proof. The proof is similar in spirit to (IV.22). Suppose g preserves orthogonality. Let $\{e_1,\ldots,e_n\}$ be an orthogonal basis for V . Select $\{f_1,\ldots,f_n\}$ in V with $\beta(e_i,f_j) = \delta_{ij}$. Then $\{f_1,\ldots,f_n\}$ is also a basis for V . Let $\alpha_i = \beta(ge_i,gf_i)$ for $1 \leq i \leq n$. For $i \neq j$,

$$\beta(e_i + e_j,f_i - f_j) = 0$$

implies

$$\beta(ge_i + ge_j,gf_i - gf_j) = 0 .$$

Hence $\alpha_i = \alpha_j$. Let $\alpha = \alpha_i$. If x and y are in V , then

$$x = a_1e_1 + \cdots + a_ne_n ,$$
$$y = b_1f_1 + \cdots + b_nf_n$$

for suitable a_i and b_j in R. Then, a direct computation gives

$$\beta(gx,gy) = \alpha(\beta(x,y))^{\mu}.$$

The converse is immediate.

(V.43) COROLLARY. If the semi-linear isomorphism $g : V \to V$ preserves orthogonality, then

(a) $\beta(x,y) = 0$ if and only if $\beta(gx,gy) = 0$,

(b) g^{-1} preserves orthogonality, and

(c) $\beta(gx,gy) = \alpha(\beta(x,y))^{\mu}$ (α a unit)

for x and y in V.

Proof. Suppose $g : V \to V$ is a semi-linear isomorphism and the inner-automorphism induced by g is on $O(V)$, i.e., $\Phi_g(\sigma) = g\sigma g^{-1}$ is in $O(V)$ for each σ in $O(V)$. We claim g will preserve orthogonality.

To show this, suppose first that x and y are nonisotropic vectors in V satisfying $\beta(x,y) = 0$. Let σ_x and σ_y be symmetries determined by x and y. Since $\beta(x,y) = 0$, we have $\sigma_x\sigma_y = \sigma_y\sigma_x$.

Consider the symmetry σ_x. Since σ_x is an involution in $GL(V)$, $\Phi_g(\sigma_x)$ will be an involution in $GL(V)$, with negative space $g(Rx)$ and positive space $g(Rx)^{\perp}$. Since Φ_g is on $O(V)$, $\Phi_g(\sigma_x)$ is in $O(V)$ and $V = g(Rx) \perp g(Rx)^{\perp}$. Hence $\Phi_g(\sigma_x)$ is the symmetry σ_{gx} (this is computed explicitly in the proof of (V.45)). Similar observations are true about $\Phi_g(\sigma_y)$.

Since $\sigma_{gx}\sigma_{gy} = \sigma_{gy}\sigma_{gx}$, then from (V.1), either $\beta(gx,gy) = 0$ or $Rgx = Rgy$.

Suppose $Rgx = Rgy$. Then $Rx = Ry$ and $x = \alpha y$ for some α in R . But $\beta(x,y) = 0$. Thus $\beta(x,x) = \beta(x,\alpha y) = \alpha\beta(x,y) = 0$ and this contradicts the fact that x is nonisotropic. Thus $\beta(gx,gy) = 0$. Hence g preserves the orthogonality of nonisotropic vectors.

Next suppose x is nonisotropic and y is unimodular with $\beta(x,y) = 0$. Then y is in $(Rx)^{\perp}$. But x is nonisotropic. Thus $V = Rx \perp (Rx)^{\perp}$ and $(Rx)^{\perp}$ has an orthogonal basis, say $\{z_1,\ldots,z_{n-1}\}$, with each basis element nonisotropic and orthogonal to x . If $y = \sum\alpha_i z_i$ then

$$\begin{aligned}\beta(gx,gy) &= \beta(gx,\textstyle\sum\alpha_i^{\mu} gz_i)\\ &= \sum_i \alpha_i^{\mu}\beta(gx,gz_i) = 0 .\end{aligned}$$

We now complete the proof in a manner analogous to the proof of (V.42). Let $\{e_1,\ldots,e_n\}$ be an orthogonal basis for V and select a "dual" basis $\{f_1,\ldots,f_n\}$ satisfying $\beta(e_i,f_j) = \delta_{ij}$. Then, the e_i are nonisotropic and the f_i are unimodular. Let $\alpha_i = \beta(ge_i,gf_i)$. Let $i \neq j$. Two cases occur, either $\beta(e_i + e_j,e_i + e_j) =$ unit or $\beta(e_i + e_j,e_i + e_j) =$ non-unit . In the former case, using $f_i - f_j$

(which is unimodular) in a manner analogous to the proof of (V.42), we can conclude that $\alpha_i = \alpha_j$. If $\beta(e_i + e_j, e_i + e_j)$ is a non-unit, then, since the sum of a non-unit and a unit is a unit, we have $e_i + e_j + e_k$ nonisotropic for $k \neq i,j$. Then,

$$\beta(e_i + e_j + e_k, f_i - f_j) = 0$$

implies

$$\beta(g(e_i + e_j + e_k), g(f_i - f_j)) = 0 .$$

Expanding and applying the above remarks, implies $\alpha_i = \alpha_j$. Let $\alpha = \alpha_i$. If x and y are in V , then $x = a_1e_1 + \cdots + a_ne_n$ and $y = b_1f_1 + \cdots + b_nf_n$ and

$$\begin{aligned} \beta(gx,gy) &= \beta(\textstyle\sum a_i^{\mu} ge_i, \sum b_i^{\mu} gf_i) \\ &= \textstyle\sum a_i^{\mu} b_i^{\mu} \beta(ge_i, gf_i) \\ &= \alpha(\beta(x,y))^{\mu} \quad . \end{aligned}$$

Hence, if $\beta(x,y) = 0$ then $\beta(gx,gy) = 0$. We have proven the following theorem.

<u>(V.44)</u> <u>THEOREM</u>. Let $g : V \to V$ be a semi-linear isomorphism. Then Φ_g is on $O(V)$ if and only if g preserves orthogonality.

An automorphism $\Lambda : O(V) \to O(V)$ is called a <u>radial</u> <u>automorphism</u> (see (II.F) or (IV.G)) if there is a group morphism

$$\chi : O(V) \to \text{Center}(O(V))$$

satisfying

$$\Lambda(\sigma) = \chi(\sigma)\sigma$$

for all σ in $O(V)$. A given radial automorphism Λ determines χ uniquely. Thus, we denote Λ by P_χ.

The purpose of the remaining sections is to show that whenever $\Lambda : O(V) \to O(V)$ is a group automorphism there exists a group morphism $\chi : O(V) \to \text{Center}(O(V))$ and a semi-linear isomorphism $g : V \to V$ which preserves orthogonality such that

$$\Lambda = P_\chi \circ \Phi_g .$$

We first examine the question of uniqueness.

<u>(V.45)</u> <u>THEOREM</u>. Let g_1 and g_2 be semi-linear isomorphisms of V which preserve orthogonality. Let P_{χ_1} and P_{χ_2} be radial automorphisms. Then the following are equivalent.

(a) $P_{\chi_1} \circ \Phi_{g_1} = P_{\chi_2} \circ \Phi_{g_2}$,

(b) $P_{\chi_1} = P_{\chi_2}$ and $\Phi_{g_1} = \Phi_{g_2}$,

(c) $\chi_1 = \chi_2$ and $g_1 = \alpha g_2$ for some unit α in R.

Proof. It is clear that (c) implies (b) and (b) implies (a). We assume (a) and show (c). The equation in (a) may be expressed as

$$P_\chi = \Phi_g$$

where $\chi(\sigma) = \chi_1(\sigma)\chi_2(\sigma)$ and $g = g_1 g_2^{-1}$. It suffices to show $g(x) = \alpha x$ for all x in V where α is a fixed unit of R .

Let x be nonisotropic in V and σ_x be the symmetry determined by x . We consider the action of Φ_g on σ_x . Observe $\Phi_g(\sigma_x) = g\sigma_x g^{-1}$ and, if y is in V , then

$$\begin{aligned}
(g\sigma_x g^{-1})(y) &= (g\sigma_x)(g^{-1}y) \\
&= g\left[g^{-1}y - 2\,\frac{\beta(x,g^{-1}y)}{\beta(x,x)}\,x\right] \\
&= y - 2\,\frac{\beta(x,g^{-1}y)^\mu}{\beta(x,x)^\mu}\,gx \\
&= y - 2\,\frac{\beta(gx,y)}{\beta(gx,gx)}\,gx \qquad \text{(by (V.43))} \\
&= \sigma_{gx}(y)
\end{aligned}$$

Hence, $\Phi_g(\sigma_x)$ is the symmetry σ_{gx} . On the other hand, $P_\chi(\sigma_x) = \pm\sigma_x$. Further, the sign must be positive or else $P_\chi(\sigma_x)$ is not a symmetry while $\Phi_g(\sigma_x)$ is a symmetry. Hence

$$\sigma_{gx} = \Phi_g(\sigma_x) = P_\chi(\sigma_x) = \sigma_x .$$

Thus, Rgx = line of σ_{gx} = line of $\sigma_x = Rx$ and $gx = \alpha x$ for some unit α in R .

Let $\{e_1,\ldots,e_n\}$ be an orthogonal basis for V. By the above, $ge_i = \alpha_i e_i$ where α_i is a unit in R for $1 \le i \le n$. By considering sums of the e_i's, it is straightforward to show $\alpha_i = \alpha_j$ for $1 \le i,j \le n$. Letting $\alpha = \alpha_i$, we have $gx = \alpha x$ for all x in V.

(J) THE PROJECTIVITY INDUCED BY AN AUTOMORPHISM

Let $\Lambda : O(V) \to O(V)$ be a group automorphism. This section is devoted to showing Λ induces a projectivity on the projective space $P(V)$ of V. The technique we employ is based on involutions and follows Keenan's [88] adaptation of the theory to local rings from Dieudonné's treatment in [63] for orthogonal groups over fields.

Throughout this section we assume

(a) R is a local ring having 2, 3 and 5 units,

(b) V has dimension ≥ 5, and hyperbolic rank ≥ 1,

(c) $\Lambda : O(V) \to O(V)$ is a group automorphism.

Let $\{\sigma_1,\ldots,\sigma_n\}$ be elements of $O(V)$. Let

$$\Omega(\sigma_1,\ldots,\sigma_n)$$

denote the $\Omega(V)$-normal subgroup of $O(V)$ generated by $\{\sigma_1,\ldots,\sigma_n\}$.

If $O(\sigma_i) = A_i$ and $A = \sum A_i$, then, by (V.35) and the remarks after that theorem,

$$\Omega(V,A) \leq \Omega(\sigma_1,\ldots,\sigma_n) \leq O(V,A) .$$

Observe

$$\Lambda(\Omega(\sigma_1,\ldots,\sigma_n)) = \Omega(\Lambda\sigma_1,\ldots,\Lambda\sigma_n) .$$

(V.46) THEOREM. The order of σ in $O(V)$ is R if and only if the order of $\Lambda\sigma$ is R .

Proof. Observe $\Lambda\Omega(V) = \Omega(V)$. If $O(\sigma) = R$, then

$$\Omega(V) \leq \Omega(\sigma) \leq O(V) .$$

Thus,

$$\Omega(V) \leq \Omega(\Lambda\sigma) \leq O(V)$$

and $O(\Lambda\sigma) = R$. The converse follows by using Λ^{-1} .

(V.47) THEOREM. Let σ_1 and σ_2 be involutions in $O(V)$. The following are equivalent:

(a) $\lambda_m\sigma_1 \neq \pm\lambda_m\sigma_2$,

(b) $\lambda_m(\Lambda\sigma_1) \neq \pm\lambda_m(\Lambda\sigma_2)$

where $\lambda_m : O(V) \to O(V/mV)$.

<u>Proof</u>. Observe the following statements are equivalent: $\lambda_m\sigma_1 \neq \pm\lambda_m\sigma_2$; $\lambda_m(\sigma_1\sigma_2) \neq \pm I$; $O(\sigma_1\sigma_2) = R$; $O(\Lambda\sigma_1\sigma_2) = R$; $O(\Lambda\sigma_1\Lambda\sigma_2) = R$; $\lambda_m(\Lambda\sigma_1\Lambda\sigma_2) \neq \pm I$; $\lambda_m(\Lambda\sigma_1) \neq \pm\lambda_m(\Lambda\sigma_2)$.

Recall $P(V)$ denotes the projective space of V , i.e., the set of lines of V . Let $\overline{P(V)}$ denote those lines in $P(V)$ which are generated by nonisotropic vectors. A line L in $\overline{P(V)}$ is called <u>nonisotropic</u>, <u>regular</u> or <u>non-singular</u>. Thus, $\overline{P(V)}$ consists of those lines L satisfying $V = L \perp L^{\perp}$.

Suppose σ is an extremal involution in $O(V)$ and $L(\sigma)$ denotes the line of σ . Then, since σ splits V , $V = L(\sigma) \perp L(\sigma)^{\perp}$ and $L(\sigma)$ is in $\overline{P(V)}$. On the other hand, given L in $\overline{P(V)}$, then L determines an extremal involution σ in $O(V)$ having line L — this involution σ is $\pm$(symmetry with line L). This latter observation follows from the uniqueness of the orthogonal complement of L .

Employing the automorphism $\Lambda : O(V) \to O(V)$, we construct a bijection

$$\Gamma : \overline{P(V)} \to \overline{P(V)} .$$

Let L be a line in $\overline{P(V)}$. Let σ_L be an extremal involution with line L (σ_L is unique up to $\pm I$). Then, by (V.41), $\Lambda\sigma_L$ is an extremal involution. Let

$$\bar{L} = L(\Lambda\sigma_L) .$$

Define

$$\Gamma : \overline{P(V)} \to \overline{P(V)}$$

by

$$\Gamma : L \to \bar{L} ,$$

i.e., Γ carries the line L to the line of $\Lambda\sigma_L$.

The next step is to show Γ extends naturally to all of $P(V)$. This is accomplished by realizing lines in $P(V)$ as intersections of non-singular planes. Thus we pause to record some facts about non-singular planes and involutions in $O(V)$ determined by these planes.

Recall,

$$\Pi_m : V \to V/mV$$

and

$$\lambda_m : O(V) \to O(V/mV)$$

are the natural morphisms induced by the ring morphism $R \to R/m$. We will use these maps <u>frequently</u> in this section. Indeed, the use of λ_m has already occurred.

The proof of the next lemma is similar in style to (I.7) and is straightforward.

<u>(V.48)</u> <u>LEMMA</u>. Let P_1 and P_2 be planes satsifying $\Pi_m P_1 \neq \Pi_m P_2$.

If $P_1 \cap P_2$ contains a unimodular element, then $P_1 \cap P_2 = L$ a line.

Let σ be an involution in $O(V)$. If one of the proper spaces of σ is a plane then we say σ is a plane involution. If σ is a plane involution, denote its proper space which is a plane by $P(\sigma)$.

We will use the notation $P(\sigma)$ for the plane of a plane involution and the notation $P(W)$ for the lines in a subspace W of V. We hope the similar notation does not create confusion and that this remark is sufficient warning.

(V.49) LEMMA. Let σ be an involution in $O(V)$. Then σ is a plane involution if and only if $\Lambda\sigma$ is a plane involution.

Proof. Let σ be a plane involution with plane $P(\sigma) = Re_1 \perp Re_2$. Then $\sigma = \tau_1\tau_2$ where τ_i is an extremal involution with line $L(\tau_i) = Re_i$ and τ_1 and τ_2 commute. Then $\Lambda\sigma = \Lambda(\tau_1\tau_2) = (\Lambda\tau_1)(\Lambda\tau_2)$ where $\Lambda\tau_1$ and $\Lambda\tau_2$ are commuting extremal involutions. By the remarks in (II.E), $\Lambda\sigma$ is a plane involution.

We now proceed with the construction of

$$\Gamma : P(V) \to P(V) .$$

We have described the action of Γ on nonisotropic lines in the discussion before (V.48). It remains to describe the action of Γ on lines of the form Re where e is unimodular but $\beta(e,e)$ = non-unit .

Let e be unimodular, $\beta(e,e)$ = non-unit and $L = Re$. Extend e to a basis $\{e = e_1, e_2, \ldots, e_n\}$ of V . Since $\beta : V \times V \to R$ is an inner product, the determinant of $[\beta(e_i, e_j)]$ is a unit. Thus, $\beta(e_1, e_j)$ = unit for some j , $2 \leq j \leq n$. Without loss assume $j = 2$. Set

$$P_1 = Re_1 \oplus Re_2 .$$

The action of $\beta|_{P_1 \times P_1}$ is determined by

$$\begin{bmatrix} \beta(e_1,e_1) & \beta(e_1,e_2) \\ \beta(e_2,e_1) & \beta(e_2,e_2) \end{bmatrix}$$

and this matrix has a unit for its determinant. Hence P_1 is non-singular and $V = P_1 \perp P_1^{\perp}$.

Then, if $\beta(e_1,e_3)$ = unit, define

$$P_2 = Re_1 \oplus Re_3 .$$

While, if $\beta(e_1,e_3)$ = non-unit, then replace e_3 by $e_2 + e_3$. For this new choice of e_3 again let $P_2 = Re_1 \oplus Re_3$. In either case, P_2 is also non-singular and

$$Re = P_1 \cap P_2 .$$

Thus, any line in $P(V)$ may be realized as the intersection of two non-singular planes.

Let σ_1 and σ_2 be plane involutions with planes $P(\sigma_1) = P_1$ and $P(\sigma_2) = P_2$. Further, let

$$\sigma_1 = \tau_1\tau_2 ,$$
$$\sigma_2 = \psi_1\psi_2$$

where τ_1 and τ_2 (similarly ψ_1 and ψ_2) are commuting extremal involutions. Recall, we denote the proper line of an extremal involution ρ by $L(\rho)$. Thus

$$P_1 = L(\tau_1) \perp L(\tau_2)$$

and

$$P_2 = L(\psi_1) \perp L(\psi_2) .$$

However, these lines are possibly distinct from the lines above defining P_1 and P_2 .

Next shift via the automorphism Λ .

Let
$$\Sigma_i = \Lambda\sigma_i ,$$

$$T_i = \Lambda\tau_i ,$$

and $$\Psi_i = \Lambda\psi_i \quad \text{for} \quad i = 1,2 .$$

Let

$$\bar{P}_i = P(\Sigma_i) ,$$

$$Rf_i = L(T_i) ,$$

and $$Rg_i = L(\Psi_i) \quad \text{for} \quad i = 1,2 .$$

We claim $\bar{P}_1 \cap \bar{P}_2$ is a line $\bar{L}$. If so, then we will require $\Gamma : L \to \bar{L}$.

Idea:

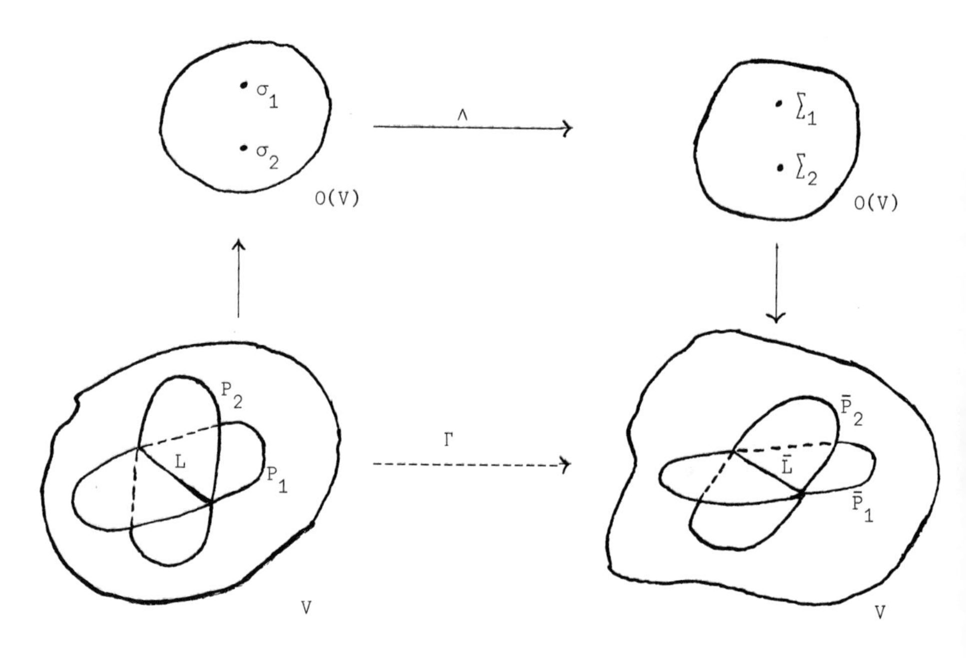

We have

$$\bar{P}_1 = Rf_1 \perp Rf_2 ,$$
$$\bar{P}_2 = Rg_1 \perp Rg_2 .$$

Since $\lambda_m \sigma_1 \neq \pm\lambda_m \sigma_2$, we have $\lambda_m \Sigma_1 \neq \pm\lambda_m \Sigma_2$ by (V.47). Hence $\Pi_m \bar{P}_1 \neq \Pi_m \bar{P}_2$. To show $\bar{P}_1 \cap \bar{P}_2$ is a line, we use (V.48) and thus show $\bar{P}_1 \cap \bar{P}_2$ has a unimodular element.

Suppose $\bar{P}_1 \cap \bar{P}_2$ does not have a unimodular element. Then

$$\Pi_m(\bar{P}_1 \cap \bar{P}_2) = 0 .$$

We will use this and arrive at a contradiction. Since $\Pi_m \bar{P}_1 \neq \Pi_m \bar{P}_2$, then $\Pi_m g_1$ or $\Pi_m g_2$ is not in $\Pi_m \bar{P}_1$. Suppose $\Pi_m g_1$ is not in $\Pi_m P_1$. Then

$$g_1 = af_1 + bf_2 + w$$

where w is in $\bar{P}_1^{\perp}$ and $\Pi_m w \neq 0$, i.e., w is unimodular. Set

$$W = Rf_1 + Rf_2 + Rw .$$

Then, it is easy to see W is a subspace of V and

$$W = Rf_1 \perp Rf_2 \perp Rw .$$

Two cases occur dependent upon the norm of w , i.e., $\beta(w,w)$.

(a) Suppose $\beta(w,w) = \text{unit}$. Then extend $\{f_1, f_2, w\}$ to an orthogonal basis for V , i.e.,

$$V = Rf_1 \perp Rf_2 \perp Rw \perp R\bar{e}_4 \perp \cdots \perp R\bar{e}_n \ .$$

(b) Suppose $\beta(w,w) = \text{non-unit}$. Select a unimodular vector $\bar{e}_4$ in $\bar{P}_1^{\perp}$ satisfying $\beta(w,\bar{e}_4) = \text{unit}$. Then complete this set of vectors to an orthogonal basis for V , i.e.,

$$V = Rf_1 \perp Rf_2 \perp (Rw \oplus R\bar{e}_4) \perp R\bar{e}_5 \perp \cdots \perp R\bar{e}_n \ .$$

Next consider the unimodular vector g_2 . Note that g_2 is not in W , for if

$$g_2 = cf_1 + df_2 + \alpha g_1$$

then

$$g_2 - \alpha g_1 \equiv c_1 f_1 + df_2 \mod mV \ ,$$

and, since $\Pi_m(\bar{P}_1 \cap \bar{P}_2) = 0$,

$$g_2 - \alpha g_1 \equiv 0 \mod mV \ .$$

Hence, $\{g_1, g_2\}$ would not be R-free. Thus, g_2 is not in W and therefore

$$g_2 = z + \sum \alpha_i \bar{e}_i$$

where z is in W and, for some j , $\alpha_j \neq 0$. Let j be maximal such that $\alpha_j \neq 0$. As above, two cases occur dependent upon the norm of w .

(a) If $\beta(w,w)$ = unit, take $\bar{e} = \bar{e}_j$.

(b) If $\beta(w,w)$ = non-unit and $j > 4$, take $\bar{e} = \bar{e}_j$. If $j = 4$ then $\alpha_4 \neq 0$ and $\alpha_5 = \cdots = \alpha_n = 0$. Select f in $\bar{P}_2$ with $\beta(f,g_1) = 0$ and $\beta(f,g_2) = 1$ — this is possible since $\bar{P}_2$ is non-singular. If f is nonisotropic, take $\bar{e} = f$. If $\beta(f,f)$ = non-unit , take $\bar{e} = f + \bar{e}_5$. (Note: we use $\dim(V) \geq 5$.)

In all cases we have selected a nonisotropic $\bar{e}$ satisfying

$$\beta(f_1,\bar{e}) = \beta(f_2,\bar{e}) = \beta(g_1,\bar{e}) = 0$$

and

$$\beta(g_2,\bar{e}) \neq 0 .$$

Let τ be an extremal involution in $O(V)$ with line $L(\Lambda\tau) = R\bar{e}$. Since $\Lambda\tau$ commutes with T_1 , T_2 and Ψ_1 , τ commutes with τ_1 , τ_2 and ψ_1 . It is easy to show if

$$U = L(\tau_1) \perp L(\tau_2) \perp L(\psi_1)$$

then

(a) $L(\tau) \perp U$,

(b) $P_1 + P_2 = U$,

(c) $\dim(U) = 3$.

But then $L(\psi_2)$ is in U. This implies $L(\tau) \perp L(\psi_2)$. Thus $L(\Lambda\tau) = R\bar{e}$ is orthogonal to $L(\psi_2) = Rg_2$. But this is impossible since $\beta(\bar{e},g_2) \neq 0$ — a contradiction. Hence

$$\Pi_m(\bar{P}_1 \cap \bar{P}_2) \neq 0 .$$

Thus, to recap, we proceed as follows (note the diagram): Given a line L in $P(V)$, select non-singular planes P_1 and P_2 in V with $L = P_1 \cap P_2$. Take plane involutions σ_1 and σ_2 in $O(V)$ with $P(\sigma_1) = P_1$ and $P(\sigma_2) = P_2$. Carry σ_1 and σ_2 by Λ to plane involutions $\Lambda(\sigma_1) = \Sigma_1$ and $\Lambda(\sigma_2) = \Sigma_2$. Let $\bar{P}_1 = P(\Sigma_1)$ and $\bar{P}_2 = P(\Sigma_2)$. By the above discussion $\bar{L} = \bar{P}_1 \cap \bar{P}_2$ is a line. Hence, define

$$\Gamma : P(V) \to P(V)$$

by

$$\Gamma : L \to \bar{L} .$$

Everything would now be fine, except that we have not shown that Γ is a well-defined map, i.e., that Γ is only a function of L and not dependent upon the planes P_1 and P_2. Since Γ is well-defined on $\overline{P(V)}$, it is only necessary to consider the case where $L = Re$ and $\beta(e,e) =$ non-unit. The key step is supplied in the next technical lemma.

(V.50) LEMMA. Let σ_1 , σ_2 and σ_3 be plane involutions in $O(V)$ satisfying

(a) if $i \neq j$ then $\lambda_m \sigma_i \neq \pm\lambda_m \sigma_j$,

(b) $P(\sigma_1) \cap P(\sigma_2) \cap P(\sigma_3) = Re$ where e is unimodular and $\beta(e,e) =$ non-unit ,

(c) $P(\sigma_1) + P(\sigma_2)$ is a non-singular subspace of dimension 3 , and

(d) $\Pi_m P(\sigma_3)$ is not in $\Pi_m(P(\sigma_1) + P(\sigma_2))$.

Then

$$P(\Lambda\sigma_1) \cap P(\Lambda\sigma_2) \cap P(\Lambda\sigma_3) = P(\Lambda\sigma_1) \cap P(\Lambda\sigma_2) .$$

Proof. We adopt the following notation for the proof. Let

$$P(\sigma_1) = Re_1 \perp Re_2 ,$$

$$P(\sigma_2) = Rf_1 \perp Rf_2 ,$$

and

$$P(\sigma_3) = Rg_1 \perp Rg_2 .$$

Let $\Lambda\sigma_i = \Sigma_i$ for $1 \leq i \leq 3$. Let

$$\Gamma Re_i = R\bar{e}_i ,$$

$$\Gamma Rf_i = R\bar{f}_i ,$$

and

$$\Gamma Rg_i = R\bar{g}_i$$

for $i = 1,2$. Observe

$$P(\Sigma_1) = R\bar{e}_1 \perp R\bar{e}_2 ,$$

$$P(\Sigma_2) = R\bar{f}_1 \perp R\bar{f}_2 ,$$

and

$$P(\Sigma_3) = R\bar{g}_1 \perp R\bar{g}_2 .$$

To proceed with the proof, observe

$$f_1 = ae_1 + be_2 + w$$

where w is in $P(\sigma_1)^{\perp}$. If $\Pi_m w = 0$ then

$$f_1 \equiv ae_1 + be_2 \mod mV$$

and thus $\Pi_m(ae_1 + be_2)$ is nonisotropic. Since pre-images of units are units, $ae_1 + be_2$ is nonisotropic. Then

$$\Pi_m Re = \Pi_m(P(\sigma_1) \cap P(\sigma_2)) = \Pi_m Rf_1$$

and e is nonisotropic — a contradiction.

Thus $\Pi_m w \neq 0$ and w is unimodular. Therefore

$$\begin{aligned} P(\sigma_1) + P(\sigma_2) &= (Re_1 \perp Re_2) \oplus Rf_1 \\ &= Re_1 \perp Re_2 \perp Rw \quad . \end{aligned}$$

By hypothesis $P(\sigma_1) + P(\sigma_2)$ is a non-singular subspace. Thus w is nonisotropic. Let $\Gamma Rw = R\bar{e}_3$ and extend $\{\bar{e}_1, \bar{e}_2, \bar{e}_3\}$ to an orthogonal basis for V . Since $P(\sigma_2) \subset Re_1 \perp Re_2 \perp Rw$, we have $Rf_i \perp (Re_1 \perp Re_2 \perp Rw)^{\perp}$. Consequently, $R\bar{f}_i \perp (R\bar{e}_1 \perp R\bar{e}_2 \perp R\bar{e}_3)^{\perp}$. That is,

$$P(\textstyle\sum_2) \subset R\bar{e}_1 \perp R\bar{e}_2 \perp R\bar{e}_3 \quad .$$

Thus

$$P(\textstyle\sum_1) + P(\textstyle\sum_2) \subset R\bar{e}_1 \perp R\bar{e}_2 \perp R\bar{e}_3 \ .$$

By methods similar to the discussion preceding the theorem, it is easy to see $P(\Sigma_1) + P(\Sigma_2)$ is a subspace of dimension 3 . Hence

$$P(\Sigma_1) + P(\Sigma_2) = R\bar{e}_1 \perp R\bar{e}_2 \perp R\bar{e}_3 \ .$$

Consider the third plane $P(\sigma_3)$. By arguments similar to those already employed, it is straightforward to show that

$$\bar{g}_1 = \alpha\bar{e}_1 + \delta\bar{e}_2 + \mu\bar{e}_3 + \bar{w}$$

where $\bar{w}$ is in $(P(\Sigma_1) + P(\Sigma_2))^{\perp}$ and $\Pi_m\bar{w} \neq 0$, i.e., $\bar{w}$ is also unimodular. Reduction modulo mV implies

$$\dim \Pi_m(P(\Sigma_1) + P(\Sigma_2) + P(\Sigma_3)) \geq 4$$

in V/mV . However, the discussion in this section indicates the spaces modulo mV pair-wise intersect. Hence

$$\dim \Pi_m(P(\Sigma_1) + P(\Sigma_2) + P(\Sigma_3)) \leq 4 \ .$$

Further, for $i \neq j$, $\Pi_m P(\Sigma_i) \neq \Pi_m P(\Sigma_j)$ and $P(\Sigma_i) \cap P(\Sigma_j)$ is a line. Thus, it is easy to see

(a) $P(\Sigma_1) + P(\Sigma_2) + P(\Sigma_3)$ is a subspace of V of dimension 4 ,

(b) $P(\Sigma_1) \cap P(\Sigma_2) \cap P(\Sigma_3) = P(\Sigma_1) \cap P(\Sigma_2)$.

This finishes the proof.

(V.51) THEOREM. The mapping $\Gamma : P(V) \to P(V)$ described in this section is well-defined.

Proof. Let $L = Re_1$ be a line in $P(V)$ where $\beta(e_1,e_1)$ = non-unit. Select f in V with $\beta(e_1,f) = 1$ and consider

$$e_2 = f - \frac{1}{2}\beta(f,f)e_1 .$$

Then

(a) f is unimodular,

(b) $\beta(e_1,e_2) = \beta(e_1,f) - \frac{1}{2}\beta(f,f)\beta(e_1,e_1)$
$= 1 - \text{non-unit}$
$= \text{unit}$,

(c) $\beta(e_2,e_2) = \beta(f - \frac{1}{2}\beta(f,f)e_1, f - \frac{1}{2}\beta(f,f)e_1)$
$= \beta(f,f) - \beta(f,f)\beta(f,e_1) + \frac{1}{4}\beta(f,f)^2\beta(e_1,e_1)$
$= \frac{1}{4}\beta(f,f)^2\beta(e_1,e_1)$
$= \text{non-unit}$.

Hence, $P = Re_1 \oplus Re_2$ is a non-singular plane since

$$\det \begin{bmatrix} \beta(e_1,e_1) & \beta(e_1,e_2) \\ \beta(e_2,e_1) & \beta(e_2,e_2) \end{bmatrix} = \text{unit} .$$

Let

$$V = (Re_1 \oplus Re_2) \perp Re_3 \perp \cdots \perp Re_n .$$

Define plane involutions σ_i , $1 \leq i \leq 4$, by

$$P(\sigma_1) = Re_1 \oplus Re_2 ,$$
$$P(\sigma_2) = Re_1 \oplus R(e_2 + e_3) ,$$
$$P(\sigma_3) = Re_1 \oplus R(e_2 + e_4) ,$$

and $$P(\sigma_4) = Re_1 \oplus R(e_2 + e_3 + e_4) .$$

Let $\Sigma_i = \Lambda\sigma_i$ for $1 \leq i \leq 4$. Then, by (V.49),

$$\begin{aligned} P(\Sigma_1) \cap P(\Sigma_2) \cap P(\Sigma_3) &= P(\Sigma_1) \cap P(\Sigma_2) \\ &= P(\Sigma_1) \cap P(\Sigma_3) \\ &= P(\Sigma_2) \cap P(\Sigma_3) \end{aligned}$$

and

$$P(\Sigma_1) \cap P(\Sigma_2) \cap P(\Sigma_4) = P(\Sigma_1) \cap P(\Sigma_2) \text{ , etc.}$$

Thus

$$\bigcap_{i=1}^{4} P(\Sigma_i) = P(\Sigma_j) \cap P(\Sigma_k)$$

for $1 \leq j \neq k \leq 4$.

Let σ be any plane involution having Re_1 in $P(\sigma)$. Let $\Sigma = \Lambda\sigma$. We claim

$$P(\Sigma_1) \cap P(\Sigma_2) \subset P(\Sigma) .$$

Case (a). $\Pi_m P(\sigma) \neq \Pi_m P(\sigma_1)$. Then, either $\Pi_m P(\sigma)$ is not in $\Pi_m P(\sigma_1) + \Pi_m P(\sigma_2)$ or $\Pi_m P(\sigma)$ is not in $\Pi_m P(\sigma_1) + \Pi_m P(\sigma_3)$. So, either $P(\Sigma_1) \cap P(\Sigma_2) \cap P(\Sigma) = P(\Sigma_1) \cap P(\Sigma_2)$ or $P(\Sigma_1) \cap P(\Sigma_3) \cap P(\Sigma) = P(\Sigma_1) \cap P(\Sigma_3)$ by (V.49). But $P(\Sigma_1) \cap P(\Sigma_2) = P(\Sigma_1) \cap P(\Sigma_3)$.

Case (b). $\Pi_m P(\sigma) = \Pi_m P(\sigma_1)$. Then $\Pi_m P(\sigma)$ is not in $\Pi_m P(\sigma_2) + \Pi_m P(\sigma_4)$. Observe $P(\sigma_2) + P(\sigma_4)$ is non-singular. Thus, by (V.50),

$P(\sum_2) \cap P(\sum_4) \cap P(\sum) = P(\sum_2) \cap P(\sum_4)$. But $P(\sum_2) \cap P(\sum_4) = P(\sum_1) \cap P(\sum_2)$.

This finishes the proof of the theorem and $\Gamma : P(V) \to P(V)$ is a well-defined map. Since Λ is a group automorphism, Λ^{-1} induces a map $\bar{\Gamma} : P(V) \to P(V)$ satisfying $\Gamma\bar{\Gamma}$ = identity and $\bar{\Gamma}\Gamma$ = identity . Thus $\Gamma : P(V) \to P(V)$ is bijective.

We wish to show $\Gamma : P(V) \to P(V)$ is a projectivity. Preliminary to this, we record some properties of Γ .

(V.52) THEOREM. Let L_1 and L_2 be lines.

(a) $L_1 + L_2$ is a plane if and only if $\Gamma L_1 + \Gamma L_2$ is a plane.

(b) Let $L_1 + L_2$ be a plane. Then $L_1 \perp L_2$ if and only if $\Gamma L_1 \perp \Gamma L_2$.

Proof. Observe $L_1 + L_2$ is a plane if and only if $\Pi_m L_1 \neq \Pi_m L_2$.

Let $L_1 = Re$ and $L_2 = Rf$. The proof of (a) and (b) will be given in various cases which depend on the norm of e and f . Let $\Gamma L_1 = R\bar{e}$ and $\Gamma L_2 = R\bar{f}$. We begin a proof of (a). Assume $L_1 + L_2$ is a plane.

(1) Suppose $\beta(e,e)$ = unit. Then $\bar{e}$ is also nonisotropic. Suppose $\Pi_m \Gamma L_1 = \Pi_m \Gamma L_2$. Then $\bar{f}$ is nonisotropic and, consequently, f is nonisotropic. Let σ_1 and σ_2 be extremal involutions with lines

$L(\Lambda\sigma_1) = \Gamma L_1$ and $L(\Lambda\sigma_2) = \Gamma L_2$. Then $\lambda_m \Lambda\sigma_1 = \pm\lambda_m \Lambda\sigma_2$. Hence $\lambda_m \sigma_1 = \pm\lambda_m \sigma_2$. Thus $\Pi_m L_1 = \Pi_m L_2$ — a contradiction since $L_1 + L_2$ is a plane.

(2) Suppose $\beta(e,e)$ and $\beta(f,f)$ are non-units. We then have $\beta(\bar{e},\bar{e})$ and $\beta(\bar{f},\bar{f})$ non-units. Assume $\Pi_m \Gamma L_1 = \Pi_m \Gamma L_2$. Select $\bar{g}$ and $\bar{h}$ unimodular in V satisfying

(a) $\Pi_m \bar{g}$, $\Pi_m \bar{h}$ and $\Pi_m \bar{e}$ linearly independent,

(b) $\beta(\bar{g},\bar{e})$ = unit , $\beta(\bar{h},\bar{e})$ = unit .

Then $\beta(\bar{g},\bar{f})$ = unit and $\beta(\bar{h},\bar{f})$ = unit since they are congruent to units modulo m .

Consider plane involutions σ_1 , σ_2 , σ_3 and σ_4 in $O(V)$ such that their images have planes

$$P(\Lambda\sigma_1) = R\bar{e} + R\bar{g} ,$$

$$P(\Lambda\sigma_2) = R\bar{e} + R\bar{h} ,$$

$$P(\Lambda\sigma_3) = R\bar{f} + R\bar{g} ,$$

and

$$P(\Lambda\sigma_4) = R\bar{f} + R\bar{h} .$$

Then

$$\lambda_m \Lambda\sigma_1 = \pm\lambda_m \Lambda\sigma_3 ,$$

$$\lambda_m \Lambda\sigma_2 = \pm\lambda_m \Lambda\sigma_4 .$$

Thus,

$$\lambda_m \sigma_1 = \pm\lambda_m \sigma_3 ,$$

$$\lambda_m \sigma_2 = \pm\lambda_m \sigma_4 ,$$

and, consequently,

$$\Pi_m P(\sigma_1) = \Pi_m P(\sigma_3) \ , \quad \Pi_m P(\sigma_2) = \Pi_m P(\sigma_4) \ .$$

Hence

$$\begin{aligned} \Pi_m L_1 = \Pi_m \Gamma^{-1}(\Gamma L_1) &= \Pi_m(P(\sigma_1) \cap P(\sigma_2)) \\ &= \Pi_m(P(\sigma_3) \cap P(\sigma_4)) \\ &= \Pi_m \Gamma^{-1}(\Gamma L_2) \\ &= \Pi_m L_2 \end{aligned}$$

— a contradiction. The argument also holds for $\Gamma^{-1} : P(V) \to P(V)$, proving the converse of (a). This completes the proof of (a).

As above, the proof of (b) is given in several cases depending on the norms of e and f. Assume $L_1 \perp L_2$. We use the above notation.

(1) Suppose $\beta(e,e) = \text{unit}$ and $\beta(f,f) = \text{unit}$. Then the statement in (b) follows immediately from the construction of Γ.

(2) Suppose $\beta(e,e) = \text{non-unit}$ and $\beta(f,f) = \text{unit}$. Then $V = L_2 \perp L_2^{\perp}$ ($L_2 = Rf$) and, since $L_1 \perp L_2$, we have e in $L_2^{\perp}$. Select an orthogonal basis for $L_2^{\perp}$. Since $L_2^{\perp}$ is non-singular there is an element, say g, of this basis satisfying $\beta(e,g) = \text{unit}$. By replacing g by a suitable unit multiple, we may assume we have found a nonisotropic vector g with $\beta(e,g) = 1$. Define

$$h = \frac{-2}{\beta(g,g)}\, g + e \quad .$$

Then ,

(a) $\beta(e,h) =$ unit , and

(b) $\beta(h,h) = \beta(-\frac{2}{\beta(g,g)}g + e , -\frac{2}{\beta(g,g)}g + e)$

$$= \frac{4}{\beta(g,g)} - \frac{4}{\beta(g,g)} + \beta(e,e)$$

$$= \beta(e,e) .$$

Hence,

(a) $\beta(e + h,e - h) = 0$,

(b) $\beta(e \pm h,e \pm h) =$ unit ,

(c) $\beta(f,e \pm h) = 0$.

Thus, $Rf \perp R(e \pm h)$ which, by (1) above, implies $\Gamma Rf \perp \Gamma R(e \pm h)$. But, observe $Re \subset R(e + h) \perp R(e - h)$. Thus, by the construction of Γ ,

$$\Gamma Re \subset \Gamma R(e + h) \perp \Gamma R(e - h) .$$

Hence $\Gamma Re \perp \Gamma Rf$.

(3) Suppose $\beta(e,e)$ and $\beta(f,f)$ are non-units. Select g in V with $\beta(f,g) = 1$. Then $P = Rf + Rg$ is a non-singular plane. Let $\{a_3,a_4,\dots,a_n\}$ be an orthogonal basis of $P^{\perp}$. Let

$$e = \alpha_1 f + \alpha_2 g + \alpha_3 a_3 + \cdots + \alpha_n a_n .$$

Since $\beta(e,f) = 0$,

$$0 = \alpha_1\beta(f,f) + \alpha_2\beta(g,f)$$
$$= \alpha_1\beta(f,f) + \alpha_2 .$$

Thus, $\alpha_2 = -\alpha_1\beta(f,f)$ is a non-unit since $\beta(f,f)$ is a non-unit. But $\Pi_m e \neq \Pi_m f$. Hence there must be an α_j , $3 \leq j \leq n$, which is a unit. Let

$$h = \beta(\alpha_j a_j, \alpha_j a_j)^{-1} a_j .$$

Then

(a) $\beta(e,h) = 1$,

(b) $\beta(f,h) = 0$,

(c) h is nonisotropic .

Define

$$a = (-2)\beta(h,h)^{-1}h + e .$$

Then, as above,

(a) $\beta(a,e) = \text{unit}$,

(b) $\beta(a,a) = \beta(e,e)$,

(c) $\beta(f,e \pm a) = 0$,

(d) $\beta(e \pm a, e \pm a) = \text{unit}$.

Therefore

$$Rf \perp R(e \pm a) .$$

The proof now follows from case (2) above.

To complete the proof that $\Gamma : P(V) \to P(V)$ is a projectivity, we must show: If L, L_1 and L_2 are lines where $L_1 + L_2$ is a plane containing L then ΓL is contained in $\Gamma L_1 + \Gamma L_2$. This is easy if $L_1 + L_2$ is non-singular. Otherwise it is somewhat more complicated.

<u>(V.53)</u> <u>THEOREM</u>. Let L, L_1, L_2 be lines and suppose $L_1 + L_2$ is a plane. Then L is contained in $L_1 + L_2$ if and only if ΓL is contained in $\Gamma L_1 + \Gamma L_2$.

<u>Proof</u>. Assume L is contained in $L_1 + L_2$. We will show ΓL is contained in $\Gamma L_1 + \Gamma L_2$. The converse will follow from the symmetry of the arguments for Γ^{-1}.

Suppose $L_1 + L_2$ is non-singular. Let $(L_1 + L_2)^{\perp} = L_3 \perp \cdots \perp L_n$ where the L_i for $3 \le i \le n$ are nonisotropic lines. Observe
$V = (L_1 + L_2) \perp (L_3 \perp \cdots \perp L_n)$ implies $V = (\Gamma L_1 + \Gamma L_2) \perp (\Gamma L_3 \perp \cdots \perp \Gamma L_n)$.
Since L is in $L_1 + L_2$, we have $L \perp L_i$ for $3 \le i \le n$.
Hence $\Gamma L \perp \Gamma L_i$ for $3 \le i \le n$ by (V.52). Thus

$$\Gamma L \subset (\Gamma L_3 \perp \cdots \perp \Gamma L_n)^{\perp} = \Gamma L_1 + \Gamma L_2 .$$

Suppose $P = L_1 + L_2$ is singular, i.e., $\beta|_{P\times P}$ is not an inner product. Let $L_1 = Re$ and $L_2 = Rf$. As was done previously, the proof follows from cases which depend on the norms of e and f. Since the arguments are similar to those already given in this section, we will sketch the proof.

The idea is to construct <u>non-singular</u> spaces $W_1,\ldots,W_t$ ($t \geq 2$) with

$$P = W_1 \cap \cdots \cap W_t .$$

The spaces $W_1,\ldots,W_t$ determine non-singular spaces $\Gamma W_1,\ldots,\Gamma W_t$ satisfying

$$\Gamma L_1 + \Gamma L_2 = \Gamma W_1 \cap \cdots \cap \Gamma W_t .$$

Further, $L \perp W_i^{\perp}$ for $1 \leq i \leq t$. Hence, by (V.52), $\Gamma L \perp (\Gamma W_i)^{\perp}$ for $1 \leq i \leq t$.

One concludes ΓL is contained in $\Gamma W_1 \cap \cdots \cap \Gamma W_t$, i.e., ΓL is contained in $\Gamma L_1 + \Gamma L_2$. The details of this are straightforward. Hence, for each case, we only show how to select the non-singular spaces $W_1,\ldots,W_t$.

(1) Suppose $\beta(e,e) = \text{unit}$ and $\beta(f,f) = \text{non-unit}$. Then

$$\begin{bmatrix} \beta(e,e) & \beta(e,f) \\ \beta(f,e) & \beta(f,f) \end{bmatrix} \equiv \begin{bmatrix} \text{unit} & 0 \\ 0 & 0 \end{bmatrix} \bmod m$$

since $P = L_1 + L_2$ is singular. Select g in V such that $\beta(f,g) = \text{unit}$. Then the matrix of β restricted to the space $Re + Rf + Rg$ is congruent to

$$\begin{bmatrix} \text{unit} & 0 & ? \\ 0 & 0 & \text{unit} \\ ? & \text{unit} & ? \end{bmatrix}$$

modulo m . This matrix has unit determinant. Hence,

$W_1 = Re + Rf + Rg$ is non-singular. Let h be unimodular in $W_1^{\perp}$ and take $W_2 = Re + Rf + R(g + h)$. Then W_2 is non-singular and $P = W_1 \cap W_2$.

(2) Suppose $\beta(e,e) = \text{unit}$ and $\beta(f,f) = \text{unit}$. Since P is singular, we have

$$\det \begin{bmatrix} \beta(e,e) & \beta(e,f) \\ \beta(f,e) & \beta(f,f) \end{bmatrix} \equiv 0 \mod m .$$

Hence $\beta(e,e)\beta(f,f) - \beta(e,f)^2 \equiv 0 \mod m$. Since P is a plane, the hyperplanes $(Re)^{\perp}$ and $(Rf)^{\perp}$ intersect and produce a subspace of dimension $n - 2$ (e.g., see (I.7)). Thus, it is easy to see there is a g in V with $\beta(e,g) = 0$ and $\beta(f,g) = \text{unit}$. Then

$$\det \begin{bmatrix} \beta(e,e) & \beta(e,f) & 0 \\ \beta(f,e) & \beta(f,f) & \text{unit} \\ 0 & \text{unit} & \beta(g,g) \end{bmatrix} \not\equiv 0 \mod m .$$

and $Re + Rf + Rg$ is non-singular. Take $W_1 = Re + Rf + Rg$. Let h be unimodular in $W_1^{\perp}$ and let $W_2 = Re + Rf + R(g + h)$. Then W_2 is non-singular and $P = W_1 \cap W_2$.

(3) Suppose $\beta(e,e) = \text{non-unit}$ and $\beta(f,f) = \text{non-unit}$. Then, since P is singular, $\beta(e,f)$ is a non-unit.

Consider $\Pi_m e$ and $\Pi_m f$ in V/mV . Note that $\Pi_m\beta(e,e) = 0$, $\Pi_m\beta(e,f) = 0$ and $\Pi_m\beta(f,f) = 0$. Since $Re + Rf$ is a plane, $\Pi_m e$ and $\Pi_m f$ form a plane. Let

$$H_1 = (\Pi_m e)^{\perp} \quad , \quad H_2 = (\Pi_m f)^{\perp} \ .$$

Then H_1 and H_2 are distinct hyperplanes in V/mV . Thus $\dim(H_1 \cap H_2) = n - 2$. Select $\bar{e}$ and $\bar{f}$ in V satisfying

(a) $\Pi_m\bar{e}$ is in H_2 and $\beta(e,\bar{e}) \equiv 1$ modulo m .

(b) $\Pi_m\bar{f}$ is in H_1 and $\beta(f,\bar{f}) \equiv 1$ modulo m .

(See (III.14).) That is,

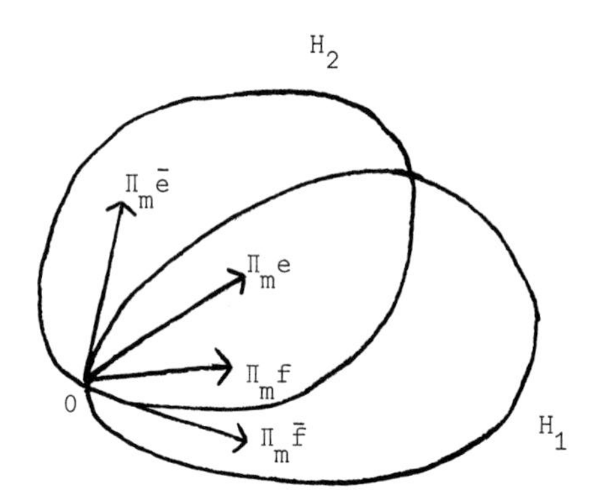

Then, it is easy to check that the matrix of $\bar{\beta}$ on V/mV restricted to the subspace spanned by $\{\Pi_m e, \Pi_m\bar{e}, \Pi_m f, \Pi_m\bar{f}\}$ has the form

$$\begin{bmatrix} 0 & 1 & 0 & 0 \\ 1 & ? & 0 & ? \\ 0 & 0 & 0 & 1 \\ 0 & ? & 1 & ? \end{bmatrix}$$

where ? represents the value of $\Pi_m\beta(\bar{e},\bar{f})$, $\Pi_m(\bar{e},\bar{e})$ and $\Pi_m(\bar{f},\bar{f})$. But this matrix has non-zero determinant. Hence, this space and its pre-image $W_1 = Re + R\bar{e} + Rf + R\bar{f}$ are non-singular. Since $\dim(V) \geq 5$, let g be unimodular in $W_1^{\perp}$. Let

$$W_2 = Re + R(\bar{e} + g) + Rf + R\bar{f}$$

and

$$W_3 = Re + R\bar{e} + Rf + R(\bar{f} + g) .$$

Then W_2 and W_3 are easily seen to be non-singular and

$$P = Re + Rf = W_1 \cap W_2 \cap W_3 .$$

This completes the proof.

The next theorem completes this section. Its proof is no more than Theorems (V.52) and (V.53). We state the theorem with the complete hypothesis of the section.

(V.54) THEOREM. Let R be a local ring having units 2, 3 and 5.[1] Let V be a symmetric inner product space of dimension ≥ 5 and hyperbolic rank ≥ 1. Let $\Lambda : O(V) \to O(V)$ be a group automorphism. Then the mapping $\Gamma : P(V) \to P(V)$ induced by Λ (as given in this section) is a projectivity.

(K) THE AUTOMORPHISMS OF THE ORTHOGONAL GROUP

We continue to employ the hypothesis of the previous section. We show in this section that if $\Lambda : O(V) \to O(V)$ is a group automorphism then there is a semi-linear isomorphism $g : V \to V$ such that Φ_g is on $O(V)$ (see (V.I)) and a radial automorphism P_χ such that

$$\Lambda = P_\chi \circ \Phi_g .$$

If $\Lambda : O(V) \to O(V)$ is a group automorphism, the previous section

---

[1] See the remarks after (V.33) concerning the units 3 and 5. These units were not used explicitly in this section and are only necessary since early in the section we utilized the normal subgroup structure of $O(V)$.

showed that Λ induced a projectivity $\Gamma : P(V) \to P(V)$. By the Fundamental Theorem of Projective Geometry (I.13) there is a semilinear isomorphism $g : V \to V$ satisfying $\Gamma L = gL$ for all lines L in $P(V)$. If $L = Re$ where e is unimodular then $gL = Rg(e)$.

(V.55) LEMMA. Let $\Lambda : O(V) \to O(V)$ be a group automorphism with induced projectivity $\Gamma : P(V) \to P(V)$. Suppose $\Gamma L = L$ for all lines L in $P(V)$. Then there is a radial automorphism P_χ (see (V.I)) such that $\Lambda = P_\chi$.

Proof. Let ρ be any element in $O(V)$. Let e be nonisotropic and consider the symmetry σ_e . Then $(\Lambda\rho)(\Lambda\sigma_e)(\Lambda\rho)^{-1}$ is an extremal involution with line $(\Lambda\rho)L$ since, by hypothesis, $\Lambda\sigma_e$ has line $L = Re$. On the other hand, $\Lambda(\rho\sigma_e\rho^{-1}) = \Lambda(\sigma_{\rho e})$ has line ρL since $\sigma_{\rho e}$ has line ρL . Thus $(\rho^{-1}\Lambda\rho)L = L$.

Let $\{e_1, e_2, \ldots, e_n\}$ be an orthogonal basis for V . Then, for $1 \le i \le n$, $(\rho^{-1}\Lambda\rho)(e_i) = \alpha_i e_i$ for α_i a unit in R since $(\rho^{-1}\Lambda\rho)Re_i = Re_i$.

Either $e_i + e_j$ is nonisotropic for given $i \neq j$ or $e_i + e_j + e_k$ is nonisotropic for $i \neq j \neq k$. Suppose $e_i + e_j$ is nonisotropic. The other case follows in a similar fashion. Then $(\rho^{-1}\Lambda\rho)(e_i + e_j) = \alpha_{ij}(e_i + e_j)$ for α_{ij} a unit. Thus

$$\begin{aligned}\alpha_{ij}(e_i + e_j) &= (\rho^{-1}\Lambda\rho)(e_i + e_j)\\ &= (\rho^{-1}\Lambda\rho)e_i + (\rho^{-1}\Lambda\rho)e_j\\ &= \alpha_i e_i + \alpha_j e_j .\end{aligned}$$

Consequently, $\alpha_{ij} = \alpha_i = \alpha_j$. We conclude that if $\alpha = \alpha_1$ then $\alpha = \alpha_i$ for $1 \le i \le n$.

Thus $(\rho^{-1}\Lambda\rho)(e_i) = \alpha e_i$ for $1 \le i \le n$ and, consequently, $(\rho^{-1}\Lambda\rho)(x) = \alpha x$ for all x in V .

Since $\rho^{-1}\Lambda\rho = \rho^{-1}(\Lambda\rho)$ is in $O(V)$,

$$\begin{aligned}v_1 = \beta(e_1,e_1) &= \beta(\rho^{-1}\Lambda\rho(e_1),\rho^{-1}\Lambda\rho(e_1))\\ &= \beta(\alpha e_1,\alpha e_1)\\ &= \alpha^2\beta(e_1,e_1)\\ &= \alpha^2 v_1 .\end{aligned}$$

Thus $\alpha = \pm 1$ and hence

$$\begin{aligned}\Lambda\rho &= \pm\rho\\ &= P_\chi(\rho)\end{aligned}$$

for a suitable $\chi : O(V) \to \mathrm{Center}(O(V))$.

The next theorem completes our treatment of the automorphisms of $O(V)$. We state the full hypothesis.

<u>(V.56)</u> <u>THEOREM</u>. (Automorphisms of $O(V)$) Let R be a local ring having 2 , 3 and 5 as units. Let V be a symmetric inner product space of dimension ≥ 5 and hyperbolic rank ≥ 1 . If $\Lambda : O(V) \to O(V)$ is a group automorphism then there exist a radial automorphism P_χ and a semi-linear isomorphism g with Φ_g on $O(V)$ such that

$$\Lambda = P_\chi \circ \Phi_g .$$

<u>Proof</u>. Let $\Gamma : P(V) \to P(V)$ be the projectivity of Section (J) induced by Λ . By (I.13) there is a semi-linear isomorphism $g : V \to V$ with $\Gamma L = gL$ for each line L in $P(V)$.

We claim that g preserves orthogonality. Let $\{e_1,\ldots,e_n\}$ be an orthogonal basis for V . By (V.52)(b), $\beta(ge_i,ge_j) = 0$ for $i \neq j$. An argument analogous to the one immediately before (V.44) will now show g preserves orthogonality. Thus Φ_g is an automorphism of $O(V)$.

Then $\Phi_g^{-1} \circ \Lambda$ is an automorphism of $O(V)$ whose induced projectivity fixes the lines of $P(V)$. By (V.55)

$$\Phi_g^{-1} \circ \Lambda = P_\chi .$$

The remainder is immediate.

The uniqueness of the above representation of Λ is given in (V.45). See also the remarks concerning the units 3 and 5 in the footnote to (V.54).

(L) DISCUSSION

This chapter summarized the basic theory of the orthogonal group over a commutative local ring. Whenever a difficulty was encountered with either the existence of units, sufficiently large dimension or existence of isotropic vectors, we chose a path of least resistance.

First and foremost was the assumption that 2 was a unit. When 2 is not a unit the theory is subtle, intricate and difficult. A detailed introduction to the necessary refinements when 2 is not necessarily a unit is given in Bass's "Unitary algebraic K-theory" [8]. Here are introduced the notions of the "unitary ring" and "unitary ideal" which appear inevitable in a serious study of normal subgroups of the orthogonal group over a ring. There is little doubt that the above paper is a significant contribution to the general theory of quadratic forms and their groups over rings.

The hypothesis that 3 and 5 are units is less serious and can usually be by-passed (as noted in the chapter) by a weaker hypothesis asserting that the local ring does not possess a residue class field of 3 or 5 elements. The point is that certain types of units are needed in the computations. The hypothesis on the residue class fields requires then a case-by-case treatment. For ease and clarity of exposition, we invoked the existence of units whenever necessary.

The hypothesis that the hyperbolic rank of the space is greater than or equal to 1 is definitely essential. As noted earlier this "linearizes" the context, permitting $\Omega(V)$ to behave in $O(V)$ as $E(V)$ behaves in $GL(V)$. The general theory of $O(V)$ over fields is not clearly understood when V contains no isotropic vectors. However, as noted below, O'Meara [109], [111] and Hahn [73], [74] have made substantial progress on the automorphism problem for fields and domains without assumptions on non-trivial hyperbolic rank.

Finally, low dimensions create added problems. There is not enough freedom of movement and, in addition, exceptional automorphisms appear.

This chapter and, indeed, the entire manuscript might be thought of as an introduction to the theory of geometric algebra over commutative rings. A currently fascinating project is the charting of the evolution of the basic theory of linear and geometric algebra over fields when the field is replaced by a commutative ring and the space by a finitely generated projective module over this ring. This should be contrasted with a similar current evolution of the classical theory over fields to subrings of the fields, i.e., domains, and lattices over these subrings. The former study appears to lie more in the province of commutative algebra and linear structures over commutative rings. The latter embraces hard and often deep number-theoretic properties of domains and their quotient fields, i.e., rings of arithmetic type.

Introductions to the theory of the orthogonal group over a field are given by Artin [1], O'Meara [22], Dieudonné [14], [63] and Eichler [16]. These sources provide much more extensive references than we include.

The normal subgroup theory for the orthogonal group over a commutative ring has emerged from two directions. First, a natural ring-theoretic progression of the structure theory from fields to local rings (as described in this chapter and in [87], [90]) and then to semi-local rings, e.g., [92]. Second, there was simultaneously a leap to arbitrary commutative (or non-commutative) rings and the imposition of side conditions necessary to achieve "stable" behavior — by "stable" behavior we mean results which approximate the theorems of this chapter. In the latter direction, the most significant work, of which we are aware, is Bass's treatment [8] mentioned above and Bak's unpublished thesis [34] (see also [35]) on the stable structure of quadratic modules. These approaches to the orthogonal group have been accompanied by a similar attack on the classification problem of symmetric inner product spaces. There are a number of recent papers describing the structure, invariants, Witt ring, etc. of symmetric inner product spaces, e.g., [31], [69], [78], [79], [93], [94], [95], [104], [127], [145].

The Clifford algebra emerges as an extremely useful algebraic structure in the study of the orthogonal group. A recent exposition of Clifford algebras over fields can be found in Lam's [18] monograph on quadratic forms. The general theory of the Clifford algebra over a commutative ring and its relation to the orthogonal group was recently published in a beautiful paper by Bass [40]. His purpose was to compute the "orthogonal Whitehead group." Bass shows the Clifford algebra of a finitely generated projective module carrying a symmetric inner product is a graded Azumaya algebra, and describes its automorphisms, in the classical sense, modulo its inner-automorphisms. He carefully constructs the spinor norm and illustrates that isometries of the orthogonal group are induced by inner-automorphisms of the Clifford algebra. We have adapted and specialized portions of his presentation to this chapter — in particular, the commutative exact diagrams illustrating the connections between the mappings and the groups.

We have not discussed the Brauer-Wall group in this chapter. The theory of the Brauer-Wall group over a field is discussed in Lam's [18] monograph on quadratic forms. The Brauer-Wall group is patterned after the construction of the Brauer group by Auslander and Goldman. This is an "enlarged" Brauer group and various inconveniences from the theory of quadratic forms are eliminated if the Clifford algebra is viewed as an element of the Brauer-Wall group instead of the Brauer group. (See Bass [40].) Clifford algebras have also incurred renewed interest in that they appear to play a unifying role in the K-theories of topology (see [29]).

As noted our approach to the normal subgroup theory follows with modest changes the approach of James [87] rather than the development of

Klingenberg [90]. Klingenberg also discusses the structure of the orthogonal group when the dimension of the space is ≤ 4. Here the Clifford algebra yields, for dimension ≤ 4, in a natural manner, isomorphisms of the special orthogonal groups into projective linear groups of two variables. Klingenberg then employs these isomorphisms to determine the normal subgroup structure in arbitrary finite dimensions.

THEOREM. (Klingenberg [90]) Let R be a local ring having 2 a unit. Let V be a symmetric inner product space of hyperbolic rank ≥ 1.

(a) If $\dim(V) = 2$ then $SO(V)$ is isomorphic to R^*.

(b) If $\dim(V) = 3$, then $SO(V)$ is isomorphic to $GL(V)/Center(GL(V))$.

(c) If $\dim(V) = 4$, then $SO(V)/Center(SO(V))$ is isomorphic to a subgroup of $GL_{\bar{R}}(V)/Center(GL_{\bar{R}}(V))$ where $\bar{R} = R[\sqrt{\alpha}]$ is a local ring generated from R by a quadratic extension by an element α in $SN(-I)$.

Klingenberg [90] also classifies the normal subgroups under the above hypothesis for dimension ≥ 5 and dimension 2 (for dimension 2 no hyperbolic rank is needed).

The approach of James allows consideration of the case where the residue class field may have characteristic 2. Details of this case are given in [87] and arguments are based on the avoidance of finite residue class fields of sizes 2, 3, 4, 5 or 9.

The automorphism theory is developed via symmetries and extremal involutions along the lines of Keenan's unpublished thesis [88].

The creation of the projectivity in Section (J) is essentially Keenan's modification of the approach of Rickart [121] and Dieudonné [63] over a field. It differs from Keenan's in that the concluding argument is modified so as to invoke the Fundamental Theorem of Projective Geometry as given in Chapter (I). To explain further — the theorem in Chapter (I) is a linear result and not related to either symplectic or orthogonal forms. However, in an analogous fashion, one can prove a symplectic "fundamental theorem" and an orthogonal "fundamental theorem." For example,

<u>THEOREM</u>. (Keenan) (Fundamental Theorem of Orthogonal Geometry) Let V be a symmetric inner product space of dimension ≥ 4. Let $\Gamma : P(V) \to P(V)$ be a bijective map satisfying:

(a) L is a nonisotropic line if and only if ΓL is a non-isotropic line.

(b) The lines L_1 and L_2 form a plane if and only if ΓL_1 and ΓL_2 form a plane.

(c) If $L_1 + L_2$ is a non-singular plane and L is a line in $L_1 + L_2$ then ΓL is in $\Gamma L_1 + \Gamma L_2$.

(d) Let $L_1 + L_2$ be a plane with L_2 nonisotropic and $L_1 \perp L_2$. Then, if L is in $L_1 + L_2$ then ΓL is in $\Gamma L_1 + \Gamma L_2$.

Then there is a semi-linear isomorphism $g : V \to V$ preserving orthogonality with $gL = \Gamma L$ for each line L in $P(V)$.

The proof of the above theorem is similar in spirit to (I.13).

Our approach to the automorphism theory by way of involutions is classical (see the discussion in (II)). It would currently be considered more appropriate to let G be a group satisfying $\Omega(V) \leq G \leq O(V)$ and ask, "If $\Lambda : G \to G$ is a group automorphism, is Λ a composition of a radial automorphism and a suitable inner-automorphism?" Unfortunately, here an involution approach fails since we cannot recognize the involutions which lie in $\Omega(V)$. Indeed, if V does not have non-trivial hyperbolic rank, then $\Omega(V)$ may have no non-trivial involutions. One is forced to abandon the involution technique. The most promising approach is O'Meara's residual space theory. However, again for local rings, one is thwarted since this theory does not apply directly to rings with zero divisors.

On the other hand, for fields and domains the residual space technique is far-reaching. In 1968, O'Meara [109] showed that if R is a field with 2 a unit and $|R| > 3$ and the dimension of the space is n where $n = 7$ or $n \geq 9$, then the automorphisms of $\Omega(V)$ have the above desired form. O'Meara replaced the involution arguments by his residual space techniques, a study of the derived chain of commutator subgroups, and commutator-centralizer arguments. The above result did not utilize an assumption on hyperbolic rank ≥ 1 ! Indeed, if hyperbolic rank is

≥ 1 , then the result also holds for $n = 8$. One would suspect that a careful study of O'Meara's arguments would allow the above to be extended to local rings. Later [111] O'Meara extended these results to local domains and Hasse domains (in global fields of characteristic 0 and p).

In 1975, Alex Hahn [73] extended O'Meara's approach to orthogonal groups over infinite domains (for finite fields the results are known). In general, the automorphisms of the orthogonal group and its congruence subgroups are of the above form. The case where the dimension of the space is 8 is interesting and difficult when the space has no isotropic vectors. For example, there exist symmetric inner product spaces over infinite fields with no isotropic vectors in dimension 8 for which the orthogonal group is finite. Obviously it is then "difficult" to create projectivities from isometries! Further, exceptional automorphisms appear in dimension 8 (see Hahn [74], [75]). Automorphism theory in dimension 8 is intimately connected with Cayley algebras and the "triviality principal."

The study of the automorphisms of the subgroups of the orthogonal group and their projective counterparts in dimension ≤ 6 is incomplete — indeed, nonexistent in the case of no hyperbolic rank. (See Hahn [74], [75].)

One should also mention the work of Borel - Tits in the expanded setting of linear algebraic groups. In particular, they have characterized the morphisms between simple algebraic groups of positive rank, i.e., hyperbolic rank ≥ 1 .

However, the current major efforts stem from O'Meara's groundbreaking paper in 1968. These techniques now await extension to arbitrary commutative rings.

There has been some work on the structure theory of $O(V)$ when the symmetric bilinear form $\beta : V \times V \to R$ is not non-singular. C. Riehm in "Orthogonal groups over the integers of a local field (II)," Amer. J. Math. 89(1967), 549-577, generalized Klingenberg's results concerning normal subgroups to the ring of integers of a local field when β has a non-unit discriminant. Chang [54] generalized the results of Klingenberg and Riehm to semi-local domains without the assumption that β is non-singular. Chang also proved generation theorems of the congruence subgroups. To our knowledge, there is no automorphism theory in this setting.

BIBLIOGRAPHY

The bibliography is given in two sections. The first part deals with books and monographs. The second part concerns, with the exception of several classic papers, papers related to geometric algebra over rings. Further references to the extensive literature over fields are provided by Dieudonné in [14] and Baumslag in [9] and [10].

A. Books and Monographs

1. Artin, E., Geometric Algebra, Wiley-Interscience, New York, 1957.
2. Baer, R., Linear Algebra and Projective Geometry, Academic Press, New York, 1952.
3. Bass, H., Lectures on Topics in Algebraic K-Theory, Tata Institute, Bombay, India, 1967.
4. ________, Algebraic K-Theory, Benjamin, New York, 1968.
5. ________, Introduction to Some Methods of Algebraic K-Theory, Regional Conf. Series #20, Amer. Math. Soc., Providence, R.I., 1973.
6. ________, (Ed.), Algebraic K-Theory, I, Lecture Notes in Math. #341, Springer-Verlag, New York-Berlin, 1973.
7. ________, (Ed.), Algebraic K-Theory, II, Lecture Notes in Math. #342, Springer-Verlag, New York-Berlin, 1973.
8. ________, (Ed.), Algebraic K-Theory, III, Lecture Notes in Math. #343, Springer-Verlag, New York-Berlin, 1973.
9. Baumslag, G., (Ed.), Reviews on Infinite Groups, Part I, Amer. Math. Soc., Providence, R.I., 1974.

10. __________, (Ed.), Reviews on Infinite Groups, Part II, Amer. Math. Soc., Providence, R.I., 1974.

11. Biggs, N., Finite Groups of Automorphisms, Cambridge Univ. Press, London, 1971.

12. Cohn, P. M., Free Rings and Their Relations, Academic Press, New York, 1971.

13. Dickson, L. E., Linear Groups with an Exposition of the Galois Field Theory, Dover Reprint, New York, 1958.

14. Dieudonné, J., La Géométrie des Groupes Classiques, 3rd Ed., Erge. der Math. und Grenz., Band 5, Springer-Verlag, New York-Berlin, 1971.

15. __________, Linear Algebra and Geometry, Houghton Mifflin, Boston, 1969.

16. Eichler, M., Quadratische Formen und Orthogonale Gruppen, Band LXIII, Die Grund. der Math. Wissenschaften, Springer-Verlag, 1952.

17. Geramita, A. V., (Ed.), Conference on Commutative Algebra, Queen's Papers #42, Queen's University, Kingston, Ontario, 1975.

18. Lam, T. Y., The Algebraic Theory of Quadratic Forms, W. A. Benjamin, Reading, Mass., 1973.

19. Milnor, J., Introduction to Algebraic K-Theory, Annals of Math. Studies, Princeton, N.J., 1971.

20. Milnor, J. and D. Husemoller, Symmetric Bilinear Forms, Ergebnisse. Math. 73, Springer-Verlag, New York-Berlin, 1973.

21. Newman, M., Integral Matrices, Academic Press, New York, 1972.

22. O'Meara, O. T., Introduction to Quadratic Forms, Band 117, Die Grund. der Math. Wissenschaften, Springer-Verlag, New York-Berlin, 1963.

23. ____________, Lectures on Linear Groups, Regional Conf. Series #22, Amer. Math. Soc., Providence, R.I., 1973.

24. Serre, J.-P., A Course in Arithmetic, Springer-Verlag, New York-Berlin, 1970.

25. Simis, A., When are Projectives Free?, Queen's Papers #21, Queen's University, Kingston, Ontario, 1969.

26. Snapper, E. and R. J. Troyer, Metric Affine Geometry, Academic Press, New York, 1971.

27. Swan, R. G., Algebraic K-Theory, Lecture Notes in Math. #76, Springer-Verlag, New York-Berlin, 1968.

B. Papers

28. Abe, E., "Chevalley groups over local rings," Tôhoku Math. J. 21(1969), 474-494.

29. Atiyah, M. F., R. Bott and A. Shapiro, "Clifford modules," Topology 3(1964), 3-38.

30. Baeza, R., "Eine Zerlegung der unitären Gruppe über lokalen Ringen," Arch. Math. 24(1973), 144-157.

31. ________, "Über die Torsion der Witt-Gruppe $W_q(A)$ eines semi-lokalen Ringes," Math. Ann. 207(1974), 121-131.

32. ________, "Eine Bemerkung über Pfisterformen," Arch. Math. 25(1974), 254-259.

33. Baeza, R. and M. Knebusch, "Anullatoren von Pfisterformen über semilokalen Ringen," preprint.

34. Bak, A., "On modules with quadratic forms," In: Algebraic K-Theory, Lecture Notes in Math. #108, Springer-Verlag, New York-Berlin, 1969.

35. _______, "The stable structure of quadratic modules," Notes, Princeton Univ., 1970.

36. Bass, H., "Big projective modules are free," Ill. J. Math. 7(1963), 24-31.

37. _______, "K-theory and stable algebra," Publ. I.H.E.S. 22 (1964), 5-60.

38. _______, "The Morita theorems," mimeographed notes.

39. _______, "Modules which support non-singular forms," J. of Algebra 13(1969), 246-252.

40. _______, "Clifford algebras and spinor norms over a commutative ring," Amer. J. Math. 96(1974), 156-206.

41. Bass, H., A. Heller and R. G. Swan, "The Whitehead group of a polynomial extension," Publ. I.H.E.S. 22(1964), 545-563.

42. Bass, H., J. Milnor and J.-P. Serre, "Solution of the congruence subgroup problem for SL_n $(n \geq 3)$ and Sp_{2n} $(n \geq 2)$," Publ. I.H.E.S. 33(1967), 421-499.

43. Beck, I., "Projective and free modules," Math. Z. 129(1972), 231-234.

44. Borel, A., "On the automorphisms of certain subgroups of semisimple Lie groups," *Proc. Bombay Coll. on Alg. Geom.* (1968), 43-73.

45. Borel, A. and J. Tits, "On 'abstract' homomorphisms of simple algebraic groups," *Proc. Bombay Coll. on Alg. Geom.* (1968), 75-82.

46. ____________________, "Homomorphismes "abstraits" de groupes algébriques semisimples," *Ann. of Math.* 97(1973), 499-571.

47. Brenner, J. L., "The linear homogeneous group," *Ann. of Math.* 39(1938), 472-493.

48. _____________, "The linear homogeneous group, II," *Ann. of Math.* 45(1944), 101-109.

49. _____________, "The linear homogeneous group, III," *Ann. of Math.* 71(1960), 210-223.

50. Cassel, D., "Rings over which projective modules are free," Ph.D. Thesis, Syracuse University, 1967.

51. Chang, C., "Harrison's Witt ring of a generalized valuation ring," *Proc. A.M.S.* 47(1975), 15-21.

52. ________, "An elementary proof of a conjecture by Harrison," *Academia Sinica* 3(1975), 43-47.

53. ________, "The structure of the symplectic group over semi-local domains," *J. of Algebra* 35(1975), 457-476.

54. ________, "Orthogonal groups over semilocal domains," *J. of Algebra* 37(1975), 137-164.

55. Cohn, P. M., "On the structure of GL_2 of a ring," Publ. I.H.E.S. 30(1966), 5-53.

56. Coleman, D. B. and J. Cunningham, "Harrison's Witt ring of a commutative ring," J. of Algebra 18(1971), 549-564.

57. ________________, "Comparing Witt rings," J. of Algebra 28(1974), 296-303.

58. ________________, "Nondegenerate higher degree forms over Dedekind domains," Proc. A.M.S. 52(1975), 65-68.

59. Connell, I. G., "Some ring theoretic Schröder-Bernstein theorems," Trans. A.M.S. 132(1968), 335-351.

60. Cooke, G., "A weakening of the Euclidean property for integral domains and applications to algebraic number theory," preprint.

61. Craven, T. C., "Stability in Witt rings," preprint.

62. Dennin, J. and D. L. McQuillan, "A note on the classical groups over semilocal rings," Proc. Roy. Irish Acad. Sect. A, 68(1969), 1-4.

63. Dieudonné, J., "On the automorphisms of classical groups," Memoirs of A.M.S. 2(1950).

64. ________, "Sur les générateurs des groupes classiques," Summa Brasil. Math. 3(1955), 149-179.

65. Dull, M. H., "Automorphisms of PSL_2 over domains with few units," J. of Algebra 27(1973), 372-379.

66. Estes, D. and J. Ohm, "Stable range in commutative rings," J. of Algebra 7(1967), 343-362.

67. Fröhlich, A. and A. M. McEvett, "Forms over rings with involution," *J. of Algebra* 12(1969), 79-104.

68. Fujisaki, G., "A note on Witt rings over local rings," *J. Fac. Sci. Univ. Tokyo* 19(1972), 403-414.

69. Fuller, K., "Density and equivalence," *J. of Algebra* 29(1974), 528-550.

70. Gabel, M., "Lower bounds on the stable range of polynomial rings," preprint.

71. Hahn, A. J., "On the homomorphisms of the integral linear groups," *Math. Ann.* 197(1972), 234-250.

72. __________, "The isomorphisms of certain subgroups of the isometry groups of reflexive spaces," *J. of Algebra* 27(1973), 205-242.

73. __________, "On the isomorphisms of the projective orthogonal groups and their congruence subgroups," *J. Reine Angew. Math.* 273(1975), 1-22.

74. __________, "Cayley algebras and the automorphisms of $PO_8'(V)$ and $P\Omega_8(V)$," preprint.

75. __________, "Octaves, triality and Aut $PO_8'(V)$," preprint.

76. Harrison, D. K., "Commutative nonassociative algebras and cubic forms," *J. of Algebra* 32(1974), 518-528.

77. ______________, "A Grothendieck ring of higher degree forms," *J. of Algebra* 35(1975), 123-138.

78. Hornix, E. A. M., "Stiefel-Whitney invariants of quadratic forms over local rings," *J. of Algebra* 26(1973), 258-279.

79. Hsia, J. S., "On the Witt ring and some arithmetical invariants of quadratic forms," J. of Number Thy. 5(1973), 339-355.

80. Hua, L. K., "On the automorphisms of the symplectic group over any field," Ann. of Math. 49(1948), 739-759.

81. Hua, L. K. and I. Reiner, "Automorphisms of the unimodular group," Trans. A.M.S. 71(1951), 331-348.

82. Hurley, J., "Some normal subgroups of elementary subgroups of Chevalley groups over rings," Amer. J. Math. 93(1971), 1059-1069.

83. __________, "Normality and terminality in the elementary subgroups of Steinberg groups over rings," Proc. A.M.S. 34(1972), 30-34.

84. __________, "Normal closures in elementary subgroups of Chevalley groups over rings," preprint.

85. James, D. and S. M. Rosenzweig, "Associated vectors in lattices over valuation rings," Amer. J. Math. 90(1968), 295-307.

86. James, D., "On the orthogonal group $O(H(A \oplus A),A)$," J. London Math. Soc. (2)4(1972), 584-586.

87. __________, "On the structure of orthogonal groups over local rings," Amer. J. Math. 95(1973), 255-265.

88. Keenan, E. M., "On the automorphisms of classical groups over local rings," Ph.D. Thesis, M.I.T., 1965.

89. Klingenberg, W., "Lineare Gruppen über lokalen Ringen," Amer. J. Math. 83(1961), 137-153.

90. ______________, "Orthogonale Gruppen über lokalen Ringen," Amer. J. Math. 83(1961), 281-320.

91. ______________, "Symplectic groups over local rings," Amer. J. Math. 85(1963), 232-240.

92. Knebusch, M., "Isometrien über semilokalen Ringen," Math. Z. 108(1969), 255-268.

93. ___________, "Runde Formen über semilokalen Ringen," Math. Ann. 193(1971), 21-34.

94. Knebusch, M., A. Rosenberg and R. Ware, "Structure of Witt rings and quotients of Abelian group rings," Amer. J. Math. 94(1972), 119-155.

95. ____________________________________, "Signatures on semilocal rings," J. of Algebra 26(1973), 208-250.

96. Lacroix, N. H. J., "Two-dimensional linear groups over local rings," Canad. J. Math. 21(1969), 106-135.

97. Lam, T. Y. and M. K. Siu, " K_0 and K_1 — an introduction to algebraic K-theory," Amer. Math. Monthly 82(1975), 329-364.

98. Landin, J. and I. Reiner, "Automorphisms of the general linear group over a principal ideal domain," Ann. of Math. (2)65(1957), 519-526.

99. Lazard, D., "Liberté des Gros Modules Projectifs," J. of Algebra 31(1974), 437-451.

100. Limaye, N. B., "Projectivities over local rings," Math. Z. 121 (1971), 175-180.

101. McDonald, B. R., "Automorphisms of $GL_n(R)$," Trans. A.M.S. 215 (1976), 145-159.

102. _______________, "Endomorphism rings of infinitely generated projective modules," to appear: *J. of Algebra*.

103. McQueen, L. and B. R. McDonald, "Automorphisms of the symplectic group over a local ring," *J. of Algebra* 30(1974), 485-495.

104. Mandelberg, K. I., "On the classification of quadratic forms over semilocal rings," *J. of Algebra* 33(1975), 463-471.

105. Martin, K., "Orthogonal groups over $R((\pi))$," *Amer. J. Math.* 95(1973), 59-79.

106. Maxwell, G., "Infinite general linear groups over rings," *Trans. A.M.S.* 151(1970), 371-375.

107. Mennicke, J. L., "Finite factor groups of the unimodular group," *Ann. of Math.* 81(1965), 31-37.

108. O'Meara, O. T., "The automorphisms of linear groups over any integral domain," *J. Reine Angew. Math.* 222(1966), 56-100.

109. _______________, "The automorphisms of the orthogonal groups $\Omega_n(V)$ over fields," *Amer. J. Math.* XC(1968), 1260-1306.

110. _______________, "The automorphisms of the standard symplectic group over any integral domain," *J. Reine Angew. Math.* 230(1968), 104-138.

111. _______________, "The automorphisms of the orthogonal groups and their congruence subgroups over arithmetic domains," *J. Reine Angew. Math.* 238(1969), 169-206.

112. _______________, "Group theoretic characterization of transvections using CDC," *Math. Z.* 110(1969), 385-394.

113. ______________, "The symplectic group and its automorphisms," to appear.

114. O'Meara, O. T. and H. Zassenhaus, "The automorphisms of the linear congruence groups over Dedekind domains," J. of Number Thy. 1(1969), 211-221.

115. Ojanguren, M. and R. Sridharen, "A note on the fundamental theorem of projective geometry," Comment Math. Helv. 44(1969), 310-315.

116. Pomfret, J. and B. R. McDonald, "Automorphisms of $GL_n(R)$, R a local ring," Trans. A.M.S. 173(1972), 379-388.

117. Reiner, I., "Automorphisms of the symplectic modular group," Trans. A.M.S. 80(1955), 35-50.

118. _________, "A new type of automorphism of the general linear group over a ring," Ann. of Math. (2)66(1957), 461-466.

119. Rickart, C. E., "Isomorphic groups of linear transformations," Amer. J. Math. 72(1950), 451-464.

120. ______________, "Isomorphic groups of linear transformations," Amer. J. Math. 73(1951), 697-716.

121. ______________, "Isomorphisms of infinite dimensional analogues of the classical groups," Bull. A.M.S. 56(1951), 435-448.

122. Riehm, C., "The structure of the symplectic group over a valuation ring," Amer. J. Math. 88(1966), 106-128.

123. ________, "The equivalence of bilinear forms," J. of Algebra 31(1974), 45-66.

124. Robertson, E. F., "A remark on the derived group of GL(R) ," Bull. London Math. Soc. 1(1969), 160-162.

125. ______________, "On certain subgroups of GL(R) ," J. of Algebra 15(1970), 293-300.

126. ______________, "Some properties of a subgroup of Sp(R) ," J. London Math. Soc. (2)4(1971), 65-78.

127. Rosenberg, A. and R. Ware, "Equivalent topological properties of the space of signatures of a semilocal ring," preprint.

128. Rosenberg, A. and D. Zelinsky, "Automorphisms of separable algebras," Pac. J. Math. 11(1961), 1109-1117.

129. Roy, A., "Cancellation of quadratic forms over commutative rings," J. of Algebra 10(1968), 286-298.

130. Sah, C. H., "Symmetric bilinear forms and quadratic forms," J. of Algebra 20(1972), 144-160.

131. _________, "Alternating and symmetric multilinear forms and Poincaré algebras," Comm. in Algebra 2(1974), 91-116.

132. Schreier, O. and B. L. Van der Waerden, "Die Automorphismen der projektiven Gruppen," Abh. Math. Sem. Univ. Hamburg 6(1928), 303-322.

133. Siegel, C. L., "Symplectic geometry," Amer. J. Math. 65(1943), 1-86.

134. Solazzi, R. E., "The automorphisms of the symplectic congruence groups," J. of Algebra 21(1972), 91-102.

135. ______________, "On the automorphisms of certain subgroups of $PGL_n(V)$," Ill. J. Math. 16(1972), 330-349.

136. ______________, "On isomorphisms between certain congruence groups," Proc. A.M.S. 35(1972), 405-410.

137. ______________, "The isomorphisms of certain congruence subgroups, II," Canad. J. Math. 25(1973), 1006-1014.

138. ______________, "The automorphisms of the unitary groups and their congruence subgroups," Ill. J. Math. 17(1973), 153-165.

139. ______________, "Four dimensional symplectic groups," preprint.

140. Stein, M., "Generators, relations and coverings of Chevalley groups over commutative rings," Amer. J. Math. 93(1971), 965-1004.

141. Swan, R. G., "Excision in algebraic K-theory," J. Pure and Appl. Alg. 1(1973), 221-252.

142. ___________, "A cancellation theorem for projective modules in the metastable range," Inventiones Math. 27(1974), 23-43.

143. Tamagawa, T., "On the structure of orthogonal groups," Amer. J. Math. 80(1958), 191-197.

144. Vaseršteĭn, L. N., "On the stabilization of the general linear group over a ring," Mat. Sb. 79(121)(1969), 405-424 = Math. USSR Sb. 8(1969), 383-400.

145. Wagner, R. C., "Some Witt cancellation theorems," Amer. J. Math. 94(1972), 206-220.

146. Wall, C. T. C., "On the classification of Hermitian forms, I. Rings of algebraic integers," *Comp. Math.* 22(1970), 425-451.

147. ______________, "On the classification of Hermitian forms, II. Semisimple rings," *Inventiones Math.* 18(1972), 119-141.

148. ______________, "On the classification of Hermitian forms, III. Complete semilocal rings," *Inventiones Math.* 19(1973), 59-71.

149. ______________, "On the classification of Hermitian forms, IV. Adele rings." *Inventiones Math.* 23(1974), 241-260.

150. Wan, C. H., "On the automorphisms of linear groups over a non-commutative Euclidean ring of characteristic 2 ," *Sci. Record* 1(1957), 5-8.

151. Wan, Z. X. and Y. X. Wang, "On the automorphisms of symplectic groups over a field of characteristic 2 ," *Sci. Sinica* 12(1963), 289-315.

152. Ware, R., "The structure of the Witt ring and quotients of Abelian group rings," preprint.

153. Ware, R. and J. Zelmanowitz, "The Jacobson radical of the endomorphism ring of a projective module," *Proc. A.M.S.* 26 (1970), 15-20.

154. Willard, E., "Properties of projective generators," *Math. Ann.* 158(1965), 352-364.

155. Yen, S.-C., "Linear groups over a ring," *Acta. Math. Sinica* 15(1965), 455-468 = *Chinese Math. Acta.* 7(1965), 163-179.

INDEX

U